New Developments in Categorical Data Analysis for the Social and Behavioral Sciences

QUANTITATIVE METHODOLOGY SERIES
George A. Marcoulides, Series Editor

The purpose of this series is to present methodological techniques to investigators and students from all functional areas of business, although individuals from other disciplines will also find the series useful. Each volume in the series will focus on a specific method (e.g., Data Envelopment Analysis, Factor Analysis, Multilevel Analysis, Structural Equation Modeling). The goal is to provide an understanding and working knowledge of each method with a minimum of mathematical derivations.

Proposals are invited from all interested authors. Each proposal should consist of the following: (i) a brief description of the volume's focus and intended market, (ii) a table of contents with an outline of each chapter, and (iii) a curriculum vita. Materials may be sent to Dr. George A. Marcoulides, Department of Management Science, California State University, Fullerton, CA 92634.

Marcoulides • Modern Methods for Business Research

Duncan/Duncan/Strycker/Li/Alpert • An Introduction to Latent Variable Growth Curve Modeling: Concepts, Issues, and Applications

Heck/Thomas • An Introduction to Multilevel Modeling Techniques

Marcoulides/Moustaki • Latent Variable and Latent Structure Models

Hox • Multilevel Analysis: Techniques and Applications

Heck • Studying Educational and Social Policy: Theoretical Concepts and Research Methods

Van der Ark/Croon/Sijtsma • New Developments in Categorical Data Analysis for the Social and Behavioral Sciences

New Developments in Categorical Data Analysis for the Social and Behavioral Sciences

Edited by

L. Andries van der Ark
Marcel A. Croon
Klaas Sijtsma
Tilburg University

Psychology Press
Taylor & Francis Group

New York   London

Camera ready copy for this book was provided by the editors.

First published by
Lawrence Erlbaum Associates, Inc., Publishers
10 Industrial Avenue
Mahwah, New Jersey 07430

First issued in paperback 2012

This edition published 2012 by Psychology Press

Psychology Press   Psychology Press
Taylor & Francis Group   Taylor & Francis Group
711 Third Avenue   27 Church Road
New York, NY 10017   Hove, East Sussex BN3 2FA

Psychology Press is an imprint of Taylor and Francis, an informa group company

Cover design by Kathryn Houghtaling Lacey

Library of Congress Cataloging-in-Publication Data

New developments in categorical data analysis for the social and behavioral sciences / edited by L. Andries van der Ark, Marcel A. Croon, Klaas Sijtsma.
    p. cm.
Includes bibliographical references and index.
ISBN 0-8058-4728-6 (alk. paper)
ISBN 978-0-415-65042-7 (Paperback)
1. Social sciences—Statistical methods. I. Ark, L. Andries van der.
  II. Croon, Marcel A. III. Sijtsma, K. (Klaas), 1955– IV. Series.
HA29.N4533 2005
300'.1'5195—dc22
                                         2004055157
                                               CIP

Contents

Preface

Almost all research in the social and behavioral sciences, and also in economic and marketing research, criminological research, and social medical research deals with the analysis of categorical data. Categorical data are quantified as either nominal or ordinal variables. This volume is a collection of up-to-date studies on modern categorical data analysis methods, emphasizing their application to relevant and interesting data sets.

Different scores on nominal variables distinguish groups. Examples known to everyone are gender, socioeconomic status, education, religion, and political persuasion. Other examples, perhaps less well known, are the type of solution strategy used by a child to solve a mental problem in an intelligence test and different educational training programs used to teach language skills to eight-year old pupils. Because nominal scores only identify groups, calculations must use this information but no more; thus, addition and multiplication of such scores lead to meaningless results.

Different scores on ordinal variables distinguish levels of interest, but differences between such numbers hold no additional information. Such scores are rank numbers or transformations of rank numbers. Examples are the ordering of types of education according to level of sophistication, the choice of most preferred politician to run for president, the preference for type of punishment in response to burglary without using violence, and the degree in which someone who recently underwent surgery rates his or her daily quality of life as expressed on an ordered rating scale.

Originally, the analysis of categorical data was restricted to counting frequencies, collecting them in cross tables, and determining the strength of the relationship between variables. Nowadays, a powerful collection of statistical methods is available that enables the researcher to exhaust his or her categorical data in ways that seemed illusory only one or two decades ago.

A prominent breakthrough in categorical data analysis is the development and use of latent variable models. This volume concentrates on two such classes of models, latent class analysis and item response theory. These methods assume latent variables to explain the relationships among observed categorical variables. Roughly, if the latent variable is also categorical the method is called latent class analysis and if it is continuous the method is called item response theory.

Latent class analysis basically yields the classification of a group of respondents according to their most likely pattern of scores on the categorical variables. Not only does this provide insight into the mechanisms producing the data, but modern latent class analysis also allows for the estimation of, for example, factor structures and regression models conditional on the latent class structure. Item response theory leads to the identification of one or more ordinal or interval scales. In psychological and educational testing these scales are used for individual measurement of abilities and personality traits. Item response theory has been extended to also deal with, for example, hierarchical data structures and cognitive theories explaining performance on tests.

These developments are truly exiting because they enable us to get so much more out of our data than was ever dreamt of before. In fact, when realizing the potential of modern days statistical machinery one is tempted to dig up all those data sets collected not-so-long ago and re-analyze them with the latent class analysis and item response theory methods we now have at our disposal. To give the reader some flavor of these methods, the focus of most contributions in this volume has been kept applied; that is, after a method is explained, the potential of the method for analyzing categorical data is illustrated by means of a real data example. The purpose is to explain methods at a level that is accessible to researchers not trained explicitly in applied statistics and then show how it can be used effectively for solving a real data problem.

We thank the following colleagues who provided critical comments on early drafts of the papers in this volume: Coen Bernaards (AMC Cancer Research Center, Denver), Jeff Douglas (University of Illinois at Urbana-Champaign), Joop Hox (Utrecht University), Brian Junker (Carnegy Mellon University), Steffen Kühnel (Georg-August-Universität Göttingen), John J. McArdle (University of Virginia), Ernesto San Martín (Pontificia Universidad Católica de Chile), Marijtje van Duijn (University of Groningen), Peter van der Heijden (Utrecht University), and Han van der Maas (University of Amsterdam). Thanks are also due to series editor George Marcoulides, the reviewers of Lawrence Erlbaum Associates for their useful comments, and Debra Riegert, Jason Planer, and Kerry Breen of Lawrence Erlbaum Associates for technically supporting us to prepare this volume.

If there is a "latent" motivation for us to prepare this volume it is to raise interest in modern categorical data analysis in such a way that readers will find it impossible to ignore these methods in their future data analysis. If we succeed in doing this of course, this is due primarily to the contributors to this volume, whom we thank for their efforts in making it a success.

—L. Andries van der Ark
Marcel A. Croon
Klaas Sijtsma
Tilburg University

About the Authors

Timo M. Bechger (`timo.bechger@citogroep.nl`) is affiliated to the Central Institute for Educational Measurement in Arnhem, the Netherlands. He has published about topics in educational measurement, structural equation modelling, and behavior genetics. His involvement in psychometrics is of a more recent nature.

Wicher P. Bergsma (`bergsma@eurandom.tue.nl`) is postdoctorate researcher at EURANDOM, the European research institute for the study of stochastic phenomena, situated at Eindhoven University in the Netherlands. His research interests include the analysis of survey data in the social sciences, the application of statistics in industry, and nonparametric measurement theory and hypothesis tests.

Jan Boom (`j.boom@fss.uu.nl`) is assistant professor in the Department of Developmental Psychology at the Faculty of Social Sciences at Utrecht University, The Netherlands. His main research interests are modelling of cognitive and moral development.

Marcel A. Croon (`m.a.croon@uvt.nl`) is an associate professor in the Department of Methodology and Statistics at the Faculty of Social Sciences at Tilburg University, The Netherlands. He teaches courses in research methodology, applied statistics, and latent structure models. His research interests are applied statistics for the social and behavioral sciences.

Karen Draney (`kdraney@uclink4.berkeley.edu`) completed her dissertation in 1995; the topic was a psychometric model for developmental stages. She is currently a research coordinator for the Berkeley Evaluation and Assessment Research Center at the University of California, Berkeley.

Jean-Paul Fox (`fox@edte.utwente.nl`) is a post-doctoral researcher in the Department of Research Methodology, Measurement, and Data Analysis at the Faculty of Behavioral Sciences at Twente University, The Netherlands. His main research interests include Bayesian statistics and educational measurement.

Jacques Hagenaars (`jacques.a.hagenaars@uvt.nl`) is full professor in Methodology of the Social Sciences and Vice-Dean of the Faculty of Social Sciences at Tilburg University. His main research interests are in the area of longitudinal survey analysis, the analysis of categorical data, and latent class models.

Herbert Hoijtink (`h.hoijtink@fss.uu.nl`) is full professor in Applied Bayesian Statistics at the department of Methodology and Statistics of Utrecht University. His main research interest is the translation of theories into statistical models and subsequent evaluation of these theories using the resulting models. The toolkit used for the evaluation usually consists of Bayesian computation methods like MCMC, posterior predictive inference, and posterior probabilities. Another research interest is the application, evaluation, and development of psychometric models.

Henk Kelderman (`h.kelderman@psy.vu.nl`) is associate professor at the department of Work and Organizational Psychology at the Vrije Universiteit, Amsterdam, The Netherlands. His primary research interests is in measurement models for multinomial and multivariate normal item responses, particularly in the invariance properties of these models, their substantive meaning and practical consequences. Furthermore, he is interested in graphical models, mixture measurement models and mixture prediction models applied to problems in selection psychology.

Martin Knott (`m.knott@lse.ac.uk`) is a Senior Lecturer in Statistics in the Department of Statistics at the London School of Economics and Political Science, England. His main interests are in latent variable models, and distribution theory.

Olav Laudy (`o.laudy@fss.uu.nl`) is a Ph.D. student in the Department of Methodology and Statistics at the Faculty of Social Sciences at Utrecht University, The Netherlands. His main research interest is the Bayesian analysis of inequality constrained log-linear/latent class models.

Jay Magidson Ph.D. (`jay@statisticalinnovations.com`) is founder and president of Statistical Innovations, a Boston based software and consulting firm. He developed the SI-CHAID, and GOLDMineR programs, and co-developed (with Jeroen Vermunt) the Latent GOLD and Latent GOLD Choice programs. His research interests include applications of advanced statistical modelling in the social sciences.

Gunter Maris (`gunter.maris@citogroep.nl`) is a senior research scientist at the Central Institute for Educational Measurement in Arnhem the Netherlands. His main research interests are the development of statistical methods and models for psychological research.

Peter Molenaar (p.c.m.molenaar@uva.nl) is a full professor of psychology in the Methodology Division of the Department of Psychology at the Faculty of Social and Behavioral Sciences at the University of Amsterdam, The Netherlands. The general theme of his work concerns the application of mathematical theories such as catastrophe theory, signal analysis, and ergodicity to solve substantive psychological issues like developmental stage transitions, cognitive information-processing, and relationships between intra-individual and inter-individual analyses of psychological processes.

Irini Moustaki (moustaki@aueb.gr) is an assistant professor in the Department of Statistics at the Athens University of Economics and Business, Greece. Her main research interests are latent variable models, missing values, and categorical data.

Klaas Sijtsma (k.sijtsma@uvt.nl) is full professor of methodology and statistics for psychological research at the Department of Methodology and Statistics, Faculty of Social and Behavioral Sciences, Tilburg University, The Netherlands. He specializes in psychometrics and applied statistics.

L. Andries van der Ark (a.vdrark@uvt.nl) is assistant professor and postdoctoral researcher in the Department of Methodology and Statistics, Tilburg University, The Netherlands. His main research interests are item response theory, latent class modelling, and psychological testing.

Peter van Rijn (p.w.vanrijn@uva.nl) is a Ph.D. student in the Methodology Division of the Department of Psychology at the Faculty of Social and Behavioral Sciences at the University of Amsterdam, The Netherlands. His research interests include psychometrics, item response theory, time series analysis, and relationships between intra-individual and inter-individual analyses of psychological processes.

Huub H.F.M. Verstralen (huub.verstralen@cito.nl) is a senior research scientist at the Central Institute for Educational Measurement in Arnhem the Netherlands. His main research interests are IRT modelling of various response formats, psychometrics and methodology in general, and program development.

Norman D. Verhelst (norman.verhelst@citogroep.nl) is senior researcher at the National Institute for Educational Research (Cito) in Arnhem, The Netherlands. His main research topic is item response theory and the link between this theory and structural equation models. As a consultant within the organization he is acquainted with several fields of methodology, statistics, and data analysis. His favorite programming language is APL.

Jeroen K. Vermunt (`j.k.vermunt@uvt.nl`) is a professor in the Department of Methodology and Statistics at Tilburg University, the Netherlands. He teaches and publishes on methodological topics such as categorical data techniques, methods for the analysis of longitudinal and event history data, latent class and finite mixture models, latent trait models, and multilevel and random-effects models. He is developer of the LEM program for categorical data analysis and co-developer (with Jay Magidson) of the Latent GOLD and Latent GOLD Choice software packages for latent class and finite mixture modelling.

Mark Wilson (`mrwilson@socrates.berkeley.edu`) is a full professor in the Graduate School of Education at the University of California, Berkeley. He teaches courses in psychometrics, assessment in education, and applied statistics. His research interests are in developing latent structure models useful in analyzing data from cognitive studies, and measurement applications across a wide range of areas, particularly from education.

Chapter 1

Statistical Models for Categorical Variables

L. Andries van der Ark, Marcel A. Croon, and Klaas Sijtsma
Tilburg University

This volume contains a collection of papers on the analysis of categorical data by means of advanced statistical methods. Most methods presented use one or more latent variables to explain the relationships among the observed categorical variables. If the latent variables are also categorical the method is called latent class analysis (LCA) and if they are continuous the method is called item response theory (IRT) or latent variable modelling.

Both LCA and IRT are used to analyze categorical data from at least two, but often many variables collected in a multidimensional contingency table. It is for this reason that this introductory chapter starts with a brief introduction into the analysis of contingency tables, and then introduces log-linear models, LCA and IRT at a conceptual level. The chapter ends with a brief outline of the contributions to this volume.

The focus of the contributions is applied; that is, after a method is explained, the potential of the method for analyzing categorical data is illustrated by means of a real data example. The editors express their hope that this volume is helpful in guiding applied researchers in the social and the behavioral sciences, and possibly in other fields (e.g., language studies, marketing, political science, social medical research) as well, to use the advanced and multi-purpose models discussed here in their own research.

1.1 Categorical Data and Analysis of Contingency Tables

Many variables collected in social and behavioral science research are categorical. Agresti (2002) distinguishes two kinds of categorical variables. Nominal variables have two or more numerical values that distinguish classes, for example, gender [men (e.g., score 0) and women (e.g., score 1)], religion [e.g., catholic (1), protestant (2), jewish (3), islamic (4)], and political persuasion [democratic (1), republican (2)]. The scores serve to distinguish group membership. Ordinal variables have numerical values that describe an ordering. Examples are level of education [low (1), intermediate (2), high (3)], preference for a brand of beer (e.g., scores $1, \ldots, 10$; a higher score indicates a stronger preference), and level of agreement with a particular statement about abortion (e.g., $0, \ldots, 4$; a higher score indicates a stronger endorsement). In these examples, the scores serve to order the respondents on the variable of interest.

Traditionally, relationships between categorical variables are studied by means of contingency tables. The simplest contingency table gives the two-dimensional layout for two variables, such as gender and political persuasion. In the example, the table has two rows (gender) and two columns (political persuasion). For a sample of respondents, the cells of the table give the number of democratic men, republican men, democratic women, and republican women. The margins of the table give the sample frequency distributions of gender and political persuasion. An example of such a simple two-way contingency table is Panel A of Table 2.6 on page 29.

The relationship between gender and political persuasion can be studied by means of several statistics. For tables of any dimension (i.e., number of variables) and any order (i.e., number of categories per variable), the chi-square statistic (denoted χ^2) can be used to test a null hypothesis of expected cell frequencies under a particular model for the data against an unspecified alternative. An example of a more general two-way contingency table is Table 5.1 on page 85. These expected frequencies can only be calculated under certain assumptions about the population. A common assumption is that the observed marginal distributions are the population distributions, which are kept fixed, and used to calculate cell frequencies expected under independence of the variables. The assumption then is that the variables' marginal distributions generated the data. If the chi-squared statistic is significant, the null hypothesis is rejected and it is inferred that there is a relationship between the variables. Other null models can be formulated, and expected cell frequencies calculated and tested against the observed cell frequencies, using the chi-squared statistic.

For some simple tables, the strength of the relationship can be expressed by association coefficients. For example, in a 2×2 table for gender (rows; scored 0, 1) and political persuasion (columns; scored 1, 2), assuming the marginal distributions fixed one can easily verify that the table has one degree of freedom. Consider the "democratic men" cell [i.e., the (0, 1) cell]. Given fixed marginals and a given sample size, N, the expected frequency of democratic men can be calculated and compared with the observed frequency. Obviously, the observed frequency can deviate from the expected frequency by being either higher or lower, and the more it deviates in either direction, the stronger the relationship. The strength of this relationship can be expressed by the ϕ coefficient. To define this coefficient, denote the cell frequencies of the gender by political persuasion table as n_{01}, n_{02}, n_{11}, and n_{12}, and the marginal frequencies of the rows as n_{0+} and n_{1+}, and of the columns as n_{+1}, and n_{+2}. The ϕ coefficient is defined as

$$\phi = \frac{n_{01}n_{12} - n_{02}n_{11}}{\sqrt{n_{0+}n_{1+}n_{+1}n_{+2}}}.$$

The dependence of ϕ on the χ^2 statistic is clear through

$$\phi = \sqrt{\frac{\chi^2}{N}}.$$

This relationship shows that, given fixed N, the higher χ^2 (i.e., the greater the discrepancy between observed and expected cell frequencies), the higher ϕ, either positive or negative (i.e., the stronger the association between gender and political persuasion). The ϕ coefficient is equal to the Pearson product-moment correlation, applied to the respondents' nominal scores on gender (0,1) and political persuasion (1,2), and thus attains values on the interval $[-1; 1]$ provided that the marginal distributions of the variables are equal. The more the marginal distributions are different, the smaller the range of ϕ.

The dependence of ϕ on the marginal distributions of the table obviously impairs its interpretation. This drawback is remedied by Mokken's (1971; see Loevinger, 1948) H coefficient. Let ϕ_{max} denote the maximum correlation given the marginals of the table; then, the H coefficient is defined as

$$H = \frac{\phi}{\phi_{max}}.$$

Division by ϕ_{max} guarantees that the maximum H is always 1.

Probably the best known association coefficient for contingency tables is the odds-ratio (e.g., Agresti, 2002, pp. 44-47). For cell frequencies denoted

n_{01}, n_{02}, n_{11}, and n_{12}, the odds ratio, denoted O, is defined as

$$O = \frac{n_{01}n_{12}}{n_{02}n_{11}}.$$

It takes values in the interval $[0, \infty)$. An odds ratio smaller than 1 indicates a negative relationship and an odds ratio greater than 1 a positive relationship. The odds ratio is not influenced by the marginal distributions of the table. All three coefficients can be generalized to two-way contingency tables of greater order.

Another example of a contingency table in which association can be determined is the following. Imagine two psychologists who independently rated children's inclination to engage in self-directed behavior. Assume that inclination is taken as the degree to which children exhibit this kind of behavior, recorded by means of a checklist, when they are observed in a playground among their peers. Assume that both psychologists observe each child for a fixed period of time, and then rate the child as either "low-level," "average," or "high-level." For N rated children, the ratings of the two psychologists can be collected in a 3×3 contingency table, with diagonal cells containing the frequencies by which they agreed, and the off-diagonal cells the frequencies by which they disagreed. The marginal distributions express each psychologist's propensity for assigning children from the population of interest to the three categories. Assuming the marginals fixed, the expected frequencies can be calculated and compared to the observed frequencies, using a chi-squared test. The degree to which the psychologists agree can be expressed by means of Cohen's κ coefficient, which is normalized to have a maximum of 1 independent of the marginals, expressing maximum agreement, a 0 value expressing independence, and a negative minimum which depends on the marginals. Cohen's κ has been generalized to more raters and different numbers of categories used per rater, and also the differential weighing of the off-diagonal cells.

The methods discussed thus far are suited especially for small tables, but may miss several interesting effects in tables based on more variables and/or more categories per variable. For example, consider a 3×4 table with level of education in the rows (low, intermediate, high) and religion in the columns (catholic, protestant, jewish, islamic). A researcher may hypothesize, in particular, that people with a low educational level are more often catholic than expected on the basis of the marginal frequencies of low educational level and catholic religion alone. Similarly, he/she may expect higher or lower frequencies elsewhere in the table, but not everywhere. The overall chi-squared statistic and the association coefficients cannot reveal such specific effects, but log-linear models can.

Conceptually, log-linear models may be compared with analysis of variance models (Stevens, 1992, p. 502). Log-linear models compare effects of rows and columns of a contingency table with a "grand mean" and are also capable of explaining deviates from marginal effects in cells by means of interaction effects. Let X and Y be two categorical variables with values $x = 1, \ldots, m_X$ and $y = 1, \ldots, m_Y$, respectively. The natural logarithm of the expected frequency in cell (x, y), denoted e_{xy}, is modelled to be the sum of a grand mean, λ, a row effect, λ_x^X, a column effect, λ_y^Y, and an interaction effect, λ_{xy}^{XY}, such that

$$\ln e_{xy} = \lambda + \lambda_x^X + \lambda_y^Y + \lambda_{xy}^{XY}.$$

Log-linear models that contain all main effect and interaction effect parameters—so-called saturated models—cannot be tested because there are no degrees of freedom left in the data. More importantly, the principle of parsimony requires models to be as simple as possible and this is realized best when the researcher defines the effects of interest before he/she starts analyzing the data. The other effects can be set to 0, comparable to what one does with the factor loadings in a confirmatory factor analysis, and the fit of the restricted model to the data can be tested using chi-squared test statistics. Also, competing models which are nested can be tested against one another. For example, for cell (x, y) nested models with and without the interaction parameter can be tested against each other. A significant result means that observed frequency is different from the expected frequency under the null model. See Wickens (1989), Hagenaars (1990), Agresti (1996, 2002), Stevens (1992), and Andreß, Hagenaars, and Kühnel (1997) for more information on log-linear models; also see Bergsma and Croon (chap. 5, this volume).

1.2   Categorical Data and Latent Class Analysis

LCA models assume that the frequency counts in a contingency table can be explained by finding an appropriate subgrouping of respondents, such that in each table corresponding to a subgroup the cell frequencies can be explained from the marginal distributions for that table. As the subgroups are not defined a priori but estimated from the data, they are considered to be latent; hence, latent class analysis.

Assume a discrete latent variable on which homogeneous classes of respondents can be distinguished, and denote this variable θ, with W classes, indexed $w = 1, \ldots, W$. Also, assume an arbitrary number, say, J, of observed categorical variables, denoted X_j, with $j = 1, \ldots, J$, and collected in a vector $\mathbf{X}$ with realization $\mathbf{x}$. The LCA model assumes independence between the observed variables given a fixed value of θ. This is known as local independence (LI), which means that

$$P(\mathbf{X} = \mathbf{x} \mid \theta = w) = \prod_{j=1}^{J} P(X_j = x_j | \theta = w).$$

Now, using the property of independent events, A and B, that $P(A \wedge B) = P(B)P(A)$, and applying LI to $P(A)$, we may write the LCA model as (Goodman, 2002; Heinen, 1996, p. 44; McCutcheon, 2002),

$$P(\mathbf{X} = \mathbf{x} \wedge \theta = w) = P(\theta = w) \prod_{j=1}^{J} P(X_j = x_j | \theta = w).$$

The probability that a randomly chosen respondent produces score pattern $\mathbf{X} = \mathbf{x}$, is

$$P(\mathbf{X} = \mathbf{x}) = \sum_{w=1}^{W} P(\theta = w) \prod_{j=1}^{J} P(X_j = x_j | \theta = w).$$

This equation shows how the LCA models the J-variate distribution of the observable variables in terms of latent class probabilities, $P(\theta = w)$, and probabilities of having particular scores x_j on observable variable X_j ($j = 1, \ldots, J$) given class membership, $P(X_j = x_j | \theta = w)$.

The class probabilities and the conditional probabilities can be estimated from the data for several choices of the number of latent classes, W. In practical data analysis, W often varies between 1 and 5. The parameter estimates for the best-fitting model are used to estimate the discrete distribution of θ, $P(\theta = w)$, with $w = 1, \ldots, W$. This distribution can be used together with the conditional probabilities, $P(X_j = x_j | \theta = w)$, to assign people to latent classes. For respondent v, this is done using probabilities $P(\theta = w | \mathbf{X}_v = \mathbf{x}_v)$, for $w = 1, \ldots, W$, after which he/she is assigned to the class that has the greatest subjective probability.

For a given number of latent classes, one thus finds a typology for a population in terms of response patterns on J variables; that is, different classes are characterized by different patterns of scores on the J observable variables. For example, a sociologist may be interested in the types of attitudes with respect to male and female role patterns, interpreted in terms of the typical answer pattern on the J items in a questionnaire. Heinen (1996, pp. 44-49) found that three classes fitted the data best. One class (45% of the respondents) represented a pro-women's lib point of view, another class

(11%) was traditional, and the third (44%) was liberal on some issues but traditional on others. Another example comes from developmental psychology, where researchers may be interested in different developmental groups. Each group may be characterized by another solution strategy for a particular cognitive problem, which reflects the cognitive stage of the group (e.g., Bouwmeester, Sijtsma, & Vermunt, 2004; Jansen & Van der Maas, 1997; Laudy, Boom, & Hoijtink, chap. 4, this volume).

It may be noted that, thus far, latent classes have been assumed to be unordered, that is, to have nominal measurement level. This leads to an unrestricted LCA model. A recent development is to put order restrictions on the conditional probabilities, $P(X_j = x_j | \theta = w)$, so as to express the assumption that there is an ordering among the latent classes, such that people in a higher latent class have a higher probability, $P(X_j = x_j | \theta = w)$, to give a particular answer to the item. This makes sense, for example, when a higher class stands for a higher reading ability and the items contain questions about a reading text, or when higher classes correspond to progressively higher levels of endorsement with abortion and the items are positively worded statements about abortion that have to be answered on a rating scale. Croon (1990, 2002) introduced these ordered latent class models, which were studied further by Hoijtink and Molenaar (1997), Vermunt (2001) Vermunt and Magidson (chap. 3, this volume), and Van Onna (2002); see Emons, Glas, Meijer, and Sijtsma (2003) for an application to the analysis of odd response patterns on sets of cognitive test items, and Laudy, Boom, and Hoijtink (chap. 4, this volume) for a application to balance-scale data. Haberman (1979) and Heinen (1996) discussed the close mathematical relationships between log-linear models and LCA models. More information on LCA models can be found in McCutcheon (1987) and Hagenaars and McCutcheon (2002).

1.3 Categorical Data and Item Response Theory

IRT models assume that the frequency counts in a J-dimensional contingency table based on the J items from a test can be explained by one ore more continuous latent variables on which the respondents are located. For one latent variable, given a fixed value, each contingency table corresponding to this value can be explained from the marginal distributions for that table. This is the assumption of local independence (LI) for IRT models.

The item scores usually are ordinal, expressing progressively higher levels of endorsement (e.g., Masters, 1982; Samejima, 1969), and sometimes nominal, as with multiple-choice items when students select one option from four or five unordered options (e.g., Bock, 1972; Thissen & Steinberg, 1997). Assume that we have Q continuous latent variables, enumerated $\theta_1, \ldots, \theta_Q$, and collected in vector $\boldsymbol{\theta}$. Then, LI is defined as

$$P(\mathbf{X} = \mathbf{x} \mid \boldsymbol{\theta}) = \prod_{j=1}^{J} P(X_j = x_j|\boldsymbol{\theta}).$$

For simplicity we assume that one latent variable, θ, suffices to explain the data structure, and that the probability density of θ is denoted $g(\theta)$. Then, the multivariate distribution of the data can be written as

$$P(\mathbf{X} = \mathbf{x}) = \int_\theta \prod_{j=1}^{J} P(X_j = x_j|\theta)g(\theta)d\theta.$$

The difference with the multivariate distribution of the data in an LCA model is that in IRT the latent variable is continuous, thus introducing an integral instead of a summation, while in LCA models the latent variable is discrete.

IRT models impose restrictions on the conditional response probabilities, $P(X_j = x_j|\theta)$. These restrictions can be orderings only (e.g., the response probability increases in the latent variable) or consist of the choice of a parametric function, such as the normal-ogive or the logistic. Once a model is chosen, it is fitted to the data. If a misfit is obtained, either the restrictions on the conditional response probabilities, $P(X_j = x_j|\theta)$, or the dimensionality of the model are changed, or items that were badly fitted by the model are removed from the analysis. Either way, the new model is fitted to the complete data set or the original model is fitted to the modified data set. When a fitted model is obtained, parameter estimates for items are used to calibrate a scale (θ) for respondents, on which respondent v is located by means of ML (or Bayesian) estimates of θ_v ($v = 1, \ldots, N$). This scale is then used as a measurement rod for the psychological property operationalized by the items.

Many applications of IRT models exist. For example, they are used to build large item collections—item pools—with known measurement properties (Kolen & Brennan, 1995), from which tests with desirable properties can be assembled (Van der Linden, 1998). Item pools are also the basis of computerized adaptive testing, which is the one by one adminstration of items to individuals where the choice of the next item is determined by the estimate of the individual's value on the latent variable based on the previous items, until an estimate of sufficient accuracy is obtained (Van der Linden & Glas, 2000). IRT models are also used to detect items that are biased against a particular minority group (Holland & Wainer, 1993), and individuals that show atypical test performance (Meijer & Sijtsma, 2001). Another application is the study of the cognitive process underlying the item responses by means of an appropriate re-parametrization of the item parameters in an IRT model (e.g., Fischer, 1974; Embretson, 1997).

An interesting development in the 1990s has been that IRT models have become part of a larger, more encompassing statistical tool box. For example, they have become the measurement part of linear hierarchical models for analyzing nested data (Fox, chap. 12, this volume; Fox & Glas, 2001; Patz, Junker, Johnson, & Mariano, 2002). This development in IRT is comparable to the integration of multilevel models, event-history models, regression models, and factor analysis models in LCA. See Vermunt (1997) for the development of the general framework in which these models were incorporated. Both developments reflect the increased availability of advanced statistical machinery for analyzing complex data, integrating structural analysis with the analysis of differences between groups (LCA) or individual differences (IRT).

Another interesting development is the integration of LCA and IRT. For example, nonparametric IRT models often restrict the conditional probabilities, $P(X_j = x_j|\theta)$, to be nondecreasing. This assumption reflects the idea that a higher latent variable value, for example, arithmetic ability increases the probability of solving arithmetic problems correctly. This monotonicity assumption has recently inspired the approximation of continuous nonparametric IRT models by discrete ordered LCA models (Hoijtink & Molenaar, 1997; Van Onna, 2002). The idea is that a small number of ordered latent classes can approximate the continuous latent variable with sufficient accuracy, and then make available for nonparametric IRT the repertoire of standard statistical techniques needed for investigating model fit. LCA models have also been used in the context of parametric IRT models, for example, the Rasch model; see Rost (1990). Introductions to IRT models can be found in Embretson and Reise (2000), Fischer and Molenaar (1995), Van der Linden and Hambleton (1997), and Sijtsma and Molenaar (2002).

1.4 Contents of This Volume

Hagenaars (chap. 2) discusses how misclassification and measurement errors in categorical variables lead to phenomena that are similar to the well-known regression toward the mean effect for continuous variables. He argues that for categorical variables one should rather speak of tendency toward the mode and shows by means of well-chosen examples how frequently this phenomenon occurs in social science research. He also discusses how tendency toward the mode can be fixed by appropriate latent class analyses of the data.

Vermunt and Magidson (chap. 3) attempt to bridge the differences between the linear factor analysis model for continuous data and the latent class model for categorical data. In their approach, a linear approximation to the parameter estimates obtained under a particular latent class model, the latent class factor analysis model is obtained. By means of this model they ensure that the output of their analysis is similar to that of standard factor analysis, which may be easier to interpret than the output from the original LCA.

Laudy, Boom, and Hoijtink (chap. 4) use LCA to test hypotheses involving inequality restrictions (e.g., Group A is expected to perform better on test T than Group B), and discuss how a researcher may choose among competing hypotheses. The authors analyze categorical balance-task data obtained from 900 children of different age groups, and compare several theories explaining the associations in these data.

Bergsma and Croon (chap. 5) discuss a broad class of models for testing complex hypotheses about marginal distributions for categorical data. The models are defined by means of the nonlinear equality constraints imposed on the cell probabilities in the corresponding contingency table. Furthermore, the authors discuss how the maximum likelihood estimates of these constrained cell probabilities may be obtained, and how the corresponding model can be tested.

Moustaki and Knott (chap. 6) use the EM estimation procedure and a Bayesian estimation procedure to estimate the parameters of three latent variable models for categorical data. The authors demonstrate how latent variable models for categorical data can be formulated on the basis of substantial theory. They discuss the merits and the pitfalls of both estimation procedures using software that is freely available.

Van Rijn and Molenaar (chap. 7) discuss dynamic latent variable models that allow the analysis of categorical time series observations on a single subject. The authors describe a model that integrates the basic principles of the Rasch measurement model and the assumptions of a simple stochastic model for describing individual change. They also discuss parameter estimation for this dynamic Rasch model.

Van der Ark and Sijtsma (chap. 8) discuss the imputation of item scores for missing values in data stemming from the administration of tests and questionnaires. They consider simple and more complex methods for missing data handling and single and multiple imputation. They apply their methods to three real data sets in which a priori fixed numbers of scores have been deleted artificially using several missing data mechanisms, and then impute scores for the missingness thus created. The effects of each imputation method on confirmatory and exploratory IRT scale analysis is investigated.

Kelderman (chap. 9) formulates measurement models for categorical data in terms of graphical independence models, exchangeability models, and log-linear models. By bringing these concepts under a single umbrella, he demonstrates how to start from scratch and construct an IRT model using important concepts such as exchangeability and internal and external consistency as building blocks.

Bechger, Maris, Verstralen, and Verhelst (chap. 10) discuss the Nedelsky IRT model for the analysis of test data from multiple-choice items on which some of the examinees may have guessed for the correct answer. The model rests on the assumptions that an examinee first eliminates the item's distracters he or she recognizes to be incorrect, and then guesses at random from the remaining options. They apply the model to data from a national test administered to eighth grade elementary school pupils, assessing their follow-up school level.

Draney and Wilson (chap. 11) discuss the saltus IRT model. This is a mixture Rasch model. The saltus model is especially suited for developmental data stemming from several subpopulations who are in different developmental stages. The model assumes one item difficulty parameter for each item and formalizes change from one subpopulation to the next by means of a small number of parameters. The authors apply the saltus model to data obtained from 460 children ranging in age from 5 to 17 years, who took a test assessing proportional reasoning, and show how the data should be analyzed and the results interpreted.

Fox (chap. 12) discusses a multilevel IRT model. He analyzes the data of a mathematics test administered to 2196 pupils (level 1 of the multilevel IRT model) from 97 elementary schools (level 2). This chapter focuses on the goodness of fit of the multilevel IRT model and the detection of outlying response patterns.

References

Agresti, A. (1996). *An introduction to categorical data analysis.* New York: Wiley.

Agresti, A. (2002). *Categorical data analysis* (2nd ed.). New York: Wiley.

Andreß, H. J., Hagenaars, J. A. P., & Kühnel, S. (1997). *Analyse von Tabellen und kategorialen Daten* (Analysis of tables and categorical data). Berlin: Springer.

Bock, R. D. (1972). Estimating item parameters and latent ability when responses are scored in two or more nominal categories. *Psychometrika, 37*, 29-51.

Bouwmeester, S., Sijtsma, K., & Vermunt, J. K. (2004). Latent class regression analysis for describing cognitive developmental phenomena: An application to transitive reasoning. *European Journal of Developmental Psychology, 1*, 67-86.

Croon, M. A. (1990). Latent class analysis with ordered latent classes. *British Journal of Mathematical and Statistical Psychology, 43*, 171-192.

Croon, M. A. (2002). Ordering the classes. In J. A. Hagenaars & A. L. McCutcheon (Eds.), *Applied latent class analysis* (pp. 137-162). Cambridge: Cambridge University Press.

Embretson, S. E. (1997). Multicomponent response models. In W. J. van der Linden & R. K. Hambleton (Eds.), *Handbook of modern item response theory* (pp. 305-321). New York: Springer.

Embretson, S. E., & Reise, S. P. (2000). *Item response theory for psychologists.* Mahwah, NJ: Erlbaum.

Emons, W. H. M., Glas, C. A. W., Meijer, R. R., & Sijtsma, K. (2003). Person fit in order-restricted latent class models. *Applied Psychological Measurement, 27*, 459-478.

Fischer, G. H. (1974). *Einfürung in die Theorie psychologischer Tests* (Introduction to psychological test theory). Bern, Switzerland: Huber.

Fischer, G. H., & Molenaar, I. W. (Eds.). (1995). *Rasch models. Foundations, recent developments, and applications.* New York: Springer.

Fox, J. -P., & Glas, C. A. W. (2001). Bayesian estimation of a multilevel IRT model using Gibbs sampling. *Psychometrika, 66*, 271-288.

Goodman, L. A. (2002). Latent class analysis. The empirical study of latent types, latent variables, and latent structures. In J. A. Hagenaars & A. L. McCutcheon (Eds.), *Applied latent class analysis* (pp. 3-55). Cambridge: Cambridge University Press.

Haberman, S. J. (1979). *Analysis of qualitative data*, 2 vols. New York: Academic Press.

Hagenaars, J. A. P. (1990). *Categorical longitudinal data. Log-linear, panel, trend, and cohort analysis.* Newbury Park, CA: Sage.

Heinen, T. (1996). *Latent class and discrete latent trait models: Similarities and differences.* Thousand Oaks, CA: Sage.

Hoijtink, H., & Molenaar, I. W. (1997). A multidimensional item response model: Constrained latent class analysis using the Gibbs sampler and posterior predictive checks. *Psychometrika, 62*, 171-189.

Holland, P. W., & Wainer, H. (1993). *Differential item functioning.* Hillsdale, NJ: Erlbaum.

Jansen, B. R. J., & Van der Maas, H. L. J. (1997). Statistical test of the rule-assessment methodology by latent class analysis. *Developmental Review, 17*, 321-357.

Kolen, M. J., & Brennan, R. L. (1995). *Test equating. Methods and practices.* New York: Springer.

Loevinger, J. (1948). The technique of homogeneous tests compared with some aspects of 'scale analysis' and factor analysis. *Psychological Bulletin, 45*, 507-530.

Masters, G. N. (1982). A Rasch model for partial credit scoring. *Psychometrika, 47*, 149-174.

McCutcheon, A. L. (1987). *Latent class analysis.* Newbury Park, CA: Sage.

Meijer, R. R., & Sijtsma, K. (2001). Methodology review: Evaluating person fit. *Applied Psychological Measurement, 25*, 107-135.

Mokken, R. J. (1971). *A theory and procedure of scale analysis.* Berlin: De Gruyter.

Patz, R. J., Junker, B. W., Johnson, M. S., & Mariano, L. T. (2002). The hierarchical rater model for rated test items and its application to large-scale educational assessment data. *Journal of Educational and Behavioral Statistics, 27*, 341-384.

Rost, J. (1990). Rasch models in latent classes: An integration of two approaches to item analysis. *Applied Psychological Measurement, 14*, 271-282.

Samejima, F. (1969). Estimation of latent trait ability using a response pattern of graded scores. *Psychometrika Monograph, No. 17.*

Sijtsma, K., & Molenaar, I. W. (2002). *Introduction to nonparametric item response theory.* Thousand Oaks, CA: Sage.

Stevens, J. (1992). *Applied multivariate statistics for the social sciences.* Hillsdale, NJ: Erlbaum.

Thissen, D., & Steinberg, L. (1997). A response model for multiple-choice items. In W. J. van der Linden & R. K. Hambleton (Eds.), *Handbook of modern item response theory* (pp. 51-65). New York: Springer.

Van der Linden, W. J. (1998). Special issue of *Applied Psychological Measurement* on "Optimal test assembly," *22*, 195-302.

Van der Linden, W. J., & Glas, C. A. W. (2000). *Computerized adaptive testing. Theory and practice.* Dordrecht, The Netherlands: Kluwer.

Van der Linden, W. J., & Hambleton, R. K. (1997). *Handbook of modern item response theory.* New York: Springer.

Van Onna, M. J. H. (2002). Bayesian estimation and model selection in ordered latent class models for polytomous items. *Psychometrika, 67*, 519-538.

Vermunt, J. K. (1997). *Log-linear models for event histories.* Thousand Oaks, CA: Sage.

Vermunt, J. K. (2001). The use of restricted latent class models for defining and testing nonparametric and parametric item response theory models. *Applied Psychological Measurement, 25,* 283-294.

Wickens, T. D. (1989). *Multiway contingency tables analysis for the social sciences.* Hillsdale, NJ: Erlbaum.

Chapter 2

Misclassification Phenomena in Categorical Data Analysis: Regression Toward the Mean and Tendency Toward the Mode

Jacques A. Hagenaars
Tilburg University

> *"Regression toward the mean is as inevitable as death and taxes.
> ... But even more remarkable than the ubiquitousness of regression toward the mean is how commonly the phenomenon is misunderstood, usually with undesirable consequences."* (Reichardt, 1999, p. ix)

2.1 Introduction

Methodologists and statisticians have warned researchers time and again about the regression-toward-the-mean fallacy, which is essentially about confounding true changes and observed changes due to random errors (Campbell & Clayton 1961; Campbell & Kenny 1999; Desrosières 1998; Friedman 1992; Stigler 1997, 1999; Thorndike 1942; among many others). As Stigler remarks "The recurrence of regression fallacies is testimony to its subtlety, deceptive simplicity ... History suggests that this will not change soon" (Stigler 1997, p. 113).

The regression toward the mean phenomenon is usually formulated within the framework of the statistical regression model with its standard implications of linear relationships, continuous variables and normally distributed errors. However, the basic mechanisms underlying the regression artifact and its consequent deceiving appearances can be formulated and are valid outside the domain of the standard regression model. The core element in the regression artifact is the working of a random component in the observations of the dependent characteristic at the first and second time of measurement, that is, in the pre- and posttest, the before and after measurement. The error term in the statistical regression model is an instance of the random component. A probabilistic reformulation of the random term as a kind of 'misclassification' without reference to the statistical regression model is possible and provides a different but illuminating insight into regression artifacts. By means of such a reformulation it can be shown that regression toward the mean is essentially an instance of the more general tendency toward the mode and that, for instance, for U-shaped distributions 'regression toward the extremes' must be expected.

Furthermore, this reformulation enables one to see the analogous deceiving consequences of the 'regression' artifact for categorical data analysis, especially, for the analysis of turnover or transition tables. Suttcliffe and Kuha and Skinner, among many others, have presented overviews of random components, that is, unreliability and misclassifications, in categorical data (Kuha & Skinner, 1997; Suttcliffe, 1965a, 1965b). Their work can be put into a more general framework by formulating the misclassification problem within the context of a latent class model. In general, latent variable models provide an appropriate and flexible framework to deal with many kinds of 'error' in continuous and categorical data. The latent class model is the most appropriate latent variable model for categorical data and is used here. Besides the well known founding fathers of latent class analysis, Lazarsfeld, Goodman, and Haberman, many other authors have dealt with 'errors' in categorical data explicitly using the latent class framework (Bassi, Hagenaars, Croon, & Vermunt, 2000; Goodman, 1974a, 1974b; Haberman, 1974; Hagenaars, 1975, 1990, 1993, 1994, 1998; Hagenaars & McCutcheon, 2002; Harper, 1973; Langeheine & Van de Pol, 2002; Lazarsfeld, 1950; Lazarsfeld & Henry, 1969; Maccoby, 1956; Poulsen, 1982; Van de Pol & Langeheine, 1997; Vermunt, 1997; Wiggins, 1955). Several of these authors, especially, Lazarsfeld, Maccoby, and Hagenaars have also explicitly pointed to phenomena such as the frequently observed tendency for smaller groups, 'minorities' to show much more changes over time than 'majorities' and have unmasked this observed tendency as a kind of regres-

sion artifact from which no substantive conclusions may be drawn in terms
of true changes over time.

Nevertheless, the literature containing analyses of categorical data is
still flooded with misleading statements and conclusions in this respect.
By establishing more explicitly the connection between 'misclassifications'
in categorical data and the classical regression phenomenon, a better understanding may be achieved that will prevent these kinds of fallacies. The
traditional exposition of regression toward the mean using the statistical
regression model will be presented in section 2.2. It will be followed by
an explanation of the regression phenomenon in terms of 'misclassifications' and as an instance of a general tendency toward the mode. This
tendency toward the mode will be exemplified for turnover or transition
tables, along with the possibly misleading conclusions in section 2.3. The
way the tendency toward the mode works out in more complicated settings
(bigger turnover tables, turnover tables for several subgroups, data from the
nonequivalent control group design with its accompanying phenomenon of
Lord's paradox) will be discussed in section 2.4. A short discussion of the
meanings of the key terms, such as error, random component, misclassification, true scores and latent variables will conclude this chapter.

2.2 Regression Toward the Mean and Tendency Toward the Mode

2.2.1 Statistical Regression

Regression toward the mean is in a way a natural consequence of the standard regression model. Consider the simple standard regression equation
with dependent variable Y, independent variable X and error term e

$$
\begin{aligned}
Y_i &= a + b_{yx}X_i + e_i \\
&= a + r_{xy}\left(\frac{s_y}{s_x}\right)X_i + e_i.
\end{aligned}
\tag{2.1}
$$

If the standard deviations of X and Y are equal to each other: $s_x = s_y$,
then Equation 2.1 can be rewritten, in terms of x_i and y_i, the deviations
from their respective means, as

$$
\hat{y} = r_{xy}x_i.
\tag{2.2}
$$

Whenever $s_x = s_y$, then, according to Equation 2.2, if a person deviates c
X-units from the mean of X, that person's expected value on Y will deviate
less than c Y-units from the mean of Y, namely, $r_{xy} \times c$ units (assuming

without any real loss of generality that the correlation is positive and not
perfect).

If the standard deviations of X and Y are not the same, the same
phenomenon occurs, but now in terms of the standardized scores

$$\hat{z}_y = r_{xy} z_x. \qquad (2.3)$$

A person's predicted z-score on the dependent variable will always be closer
to 0, the mean z-score of Y, than that person's z-score on the independent
variable is to 0, the mean z-score of X. In that sense, in terms of z-scores,
regression toward the mean is tautological with a nonperfect correlation
coefficient, that is, with the absolute value of r being less than one.

How and under what circumstances do these simple and well-known
facts confuse so many researchers, as stated in the Introduction? Borrowing
freely from one of the earliest documented examples of the misleading effects
of regression toward the mean (McNemar, 1940), imagine a researcher that
has at her disposal the results of an IQ test administered to young orphaned
children the moment they enter the orphanage. After a stay of one year
in the orphanage, the researcher applies the same IQ test again to these
children. The correlation between the first and second measurement turned
out to be .70. Because the researcher is interested in what exactly happened
to the orphans' IQ over time, she made groups of children having the same
score on the first IQ test and computed the mean IQ on the second test for
each group. Assuming that the relationship between the first and second
measurement is linear, these mean values lie on the regression line and are
the expected values of Y_2 (IQ at time 2) regressed on Y_1 (IQ at time 1). The
outcomes are depicted in Figure 2.1, with scores on a scale from 0 to 10 and
mean 5, and show impressive evidence that being raised in a homogeneous
environment such as an orphanage leads to homogenization of intelligence.
Obviously, the more dull persons are stimulated by their environment and
have improved on the IQ test, while the comparatively bright children were
not inspired by their environment and regressed toward mediocrity. The
pattern is such that everybody moves toward the mean over time, but the
more extreme the initial score, the bigger the change. More precisely, in
regression terms given $r = .70$, Figure 2.1 shows that if a child is 5 points
away from the mean at the first measurement, the expected score of the
second measurement will be $.70 \times 5 = 3.5$ points away from the overall
mean at time 2 and will have regressed $(1 - .70) \times 5 = 1.5$ points toward
that mean; for those children that are 1 point away at time 1, the expected
deviation score will be .7, regressing .3 points, etc.

The pattern shown in Figure 2.1 occurs a lot and has led to many
(misleading) substantive conclusions in a variety of research situations. At
first sight, it is not strange that from such outcomes it has been concluded

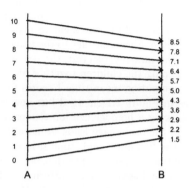

A: Scores at time 1; B: expected scores at time 2

Figure 2.1: Regression to the Mean.

that pupils that performed bad at school and therefore participated in a remedial teaching program improved after a while, especially those doing worst. The regression pattern is also implicit in many evaluation studies using time series analyses, for example, a time series of the number of street robberies in certain areas of the city. When things go worse, that is, when the number rises and seriously gets above the mean, police will start patrolling those streets leading to a decrease of the number of robberies. In more intricate ways, the regression phenomenon is connected to possibly misleading conclusions from studies interpreting substantively a negative correlation between the difference scores $(Y_2 - Y_1)$ and the initial scores Y_1 or from research designs that rely on the matching of groups with different initial scores on the dependent variable Y or from quasi-experiments like the nonequivalent control group design with its accompanying appearance of Lord's paradox (Campbell & Kenny, 1999; Lord, 1967).

But why call these regression patterns possibly misleading? After all, patrolling the streets can make the neighborhood safer, remedial teaching may help children, etcetera (as we know from randomized experiments). Let us return to our imaginary researcher. Luckily, this investigator had heard about regression artifacts and applied the two crucial tests proposed by McNemar (1940). If there really had been a homogenization effect of the orphanage environment on IQ development over time, the variance of the IQ test at time 2 should be less than the variance at time 1. However, this was not found here, but $s_1 = s_2$. Obviously, there was enough spread around the expected values of the second measurement to make the variances of time one and two equal to each other, contrary to the homogenization hypothesis. For the second test, the time reversal test, the researcher switched the roles of time 1 and 2, grouping the children together on the basis of the time 2 measurements and computing mean time 1 scores for each group. Exactly the same picture as in Figure 2.1 emerged but now with the roles of time 1 and 2 reversed showing a homogenization tendency toward the past. In other words, being in the orphanage leads to more diversity.

On the basis of the 'variance' and 'time reversal' tests, the researcher concluded that there was no real evidence that being in the orphanage leads to homogenization of intelligence. The pattern found in Figure 2.1 arose because of the random component in the model. If one selects children that have high observed scores on the first IQ test, one also selects disproportionately children with positive error terms, that just happened to be in, the (extremely) high IQ group. Our best guess upon re-measurement is that the error terms will tend to zero, causing the high IQ group as a whole to score more toward the mean the second time, if nothing else changes.

To put it a bit more formally, let us assume that Y_1, the observed IQ score at time 1, has a systematic (or true or latent) intelligence part denoted by T, and a random component e_1 (unreliability if one likes) and that the same is true for Y_2. Mainly for the sake of ease of exposition, a few simplifying assumptions will be made: the variances of the error terms are the same at both points in time, e_1 and e_2 are not correlated, and the systematic or true scores T do not change over time:

$$\begin{aligned} Y_{i1} &= T_i + e_{i1} \\ Y_{i2} &= T_i + e_{i2}. \end{aligned} \tag{2.4}$$

If the correlation in Figure 2.2 between T and Y is 0.837 ($\sqrt{.70}$) at both points in time, the correlation between the two IQ measurements will be .70 and exactly the pattern observed in Figure 2.1 will be found. Given the restrictions, the model implies equal variances and will yield the time reversal pattern, without any changes in the 'true' score T, but completely due to the random part.

2.2.2   Regression and Misclassifications

The simple model in Figure 2.2 (and Equation 2.4) can be formulated for categorical data as well. In this exposition, the categorical data conven-

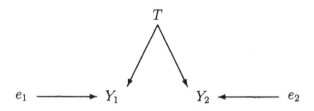

Figure 2.2: Error Model Without Latent Change.

tion will be followed where the true (latent) scores are denoted by X and
the observed pretest and posttest scores by A and B respectively. In this
subsection, all variables will be treated as categorical interval level variables. It is assumed that the true scores X do not change over time (see
Figure 2.2). The true scores X are not observed perfectly, but a random
part comes in, termed misclassifications. It is assumed that the probabilities of misclassification are the same at times one and two and that the
misclassifications at time 1 are independent of the misclassifications at time
2. Table 2.1 contains the assumed symmetrical frequency distribution of X
and the 'observed' marginal distribution of A (or B). The mean value of
X is 6 and its variance 1.27. Note further that the true categories 2, 3, 9,
and 10 are empty. The distribution of the observed scores A (or B) is not
the same as the true distribution, but resulted from the following response
probabilities. The probability that the score on X is observed exactly as
it is equals .50 (at both points in time). The probability that the observed
score is one category higher than the true score is .20, as is the probability
that the observed score is one category lower. The probability of obtaining
a score that is two categories higher (lower respectively) than the true score
is .05. From the distribution of X and the misclassification probabilities,
the joint distribution of A and B can be derived, as well as the (identical)
marginal distributions of A and B. As expected under this simple model,
the mean of A (or B) is also 6, but the variance of A (or B) is larger than
the variance of X, due to 'addition' of the random or misclassification part
(see bottom line Table 2.1). From the obtained joint distribution of A and
B (not presented here), it can be computed what the expected or mean
score on B will be for each category of A (or with the same results: what
the mean score on A will be for each category of B). The results are in Col-
umn $E(B|A)$ in Table 2.1 and show the kind of regression pattern found in

Table 2.1: Regression Toward the Mean and Tendency Toward the Mode.

	Frequency			Regression toward the mean/
	True	Observed		
Score	score X	score A, B	$E(B|A)$	Tendency toward the mode
2	0	3	4.00	+2.00
3	0	19	4.37	+1.37
4	60	68	4.71	+0.71
5	140	129	5.33	+0.33
6	200	162	6.00	0.00
7	140	129	6.67	−0.33
8	60	68	7.29	−0.71
9	0	19	7.63	−1.37
10	0	3	8.00	−2.00

$p(A = X) = .50;\ p(B = X) = .50;$
$p(\pm 1) = .20;\ p(\pm 2) = .05$ (see text);
$s_x^2 = 1.27;\ s_A^2 = s_B^2 = 2.07;\ s_X^2/s_A^2 = 0.61$

Figure 2.1: Regression toward the mean, and the more extreme a category is the more regression occurs.

This pattern of seemingly systematic change arises from the simple no-latent-change-and-independent-misclassification model just described. Table 2.2 exemplifies in detail how this happens for category $A = 5$. The first two columns of Table 2.2 present the true score distribution of X; the third column states the assumed conditional response probabilities of scoring $A = 5$, given the true score; the fourth column indicates how the composition of the group of 129 people with $A = 5$ accordingly is in terms of their true score; finally, the last column shows how each of these composing subgroups contributes to $E(B|A = 5) = 5.33$, given that under this simple model the expected observed score equals the true score. For other values of A similar tables can be constructed, leading to the outcomes in Table 2.1.

As may be obvious from this example, regression toward the mean has to do with the nature of and the assumptions about the distribution of the random term, that is, the probabilities of misclassification. Other things being equal, if the probability of misclassification increases, as a rule, given most common assumptions about the nature of misclassifications, the re-

Table 2.2: Expected Value of B, Given $A = 5$ [i.e., $E(B|A = 5)$].

True score X	Freq. X	$p(A = 5|X)$	Freq. $A = 5$	$E(B|A = 5)$
2	0	0	0	
3	0	0.05	0	
4	60	0.20	12	$12 \times 4 = \quad 48$
5	140	0.50	70	$70 \times 5 = 350$
6	200	0.20	40	$40 \times 6 = 240$
7	140	0.05	7	$7 \times 7 = \quad 49$
8	60	0	0	
9	0	0	0	
10	0	0	0	
Total			129	687

$E(B|A = 5) = 687/129 = 5.33$

gression toward the mean will increase. Furthermore, the distribution of the misclassification terms need not be symmetrical, but may favor, for example, obtaining too high scores over obtaining too low scores, giving rise to a somewhat different regression pattern. Also, in the example in Table 2.1, a certain kind of ceiling and bottom (misclassification) effects were avoided by leaving the highest and lowest categories of X empty (but existent). These and still other variants of the nature of the misclassifications, such as dependencies between the misclassifications of time 1 and 2 can easily be reckoned with and built into the (latent class) models.

As important as the distribution of the random part, is the distribution of the systematic part for the regression pattern. Because of the unimodal, symmetric distribution of X in Table 2.1, the 'mistakes' in, for example, the observed score $A = 5$ originate more from $X > 5$ than from $X < 4$; hence the downward movement toward 6. As is clear from the calculation of $E(B|A = 5)$ in Table 2.2, the tendency will be in the direction of the largest categories of X. Seen in this way, regression toward the mean is just another instance of a more general tendency toward the modal category. Given another distribution of X, a U-shaped distribution, for example, there may be regression away from the mean toward the (modal) extremes. To illustrate this possibility, the model used for the construction of Table 2.1 is applied again but now starting from a partly U-shaped distribution of X.

A different 'regression' pattern from Table 2.1 emerges in Table 2.3. Categories 4 and 7 are the modal categories of the distribution of X. The

Table 2.3: Regression From the Mean and Tendency Toward the Mode.

	Frequency			
	True	Observed		Tendency
Score	score X	score A, B	$E(B\|A)$	toward to mode
2	0	10.0	4.00	+2.00
3	0	45.0	4.11	+1.11
4	200	122.5	4.20	+0.20
5	100	105.0	4.81	−0.19
6	50	85.0	6.00	0.00
7	100	105.0	7.19	+0.19
8	200	122.5	7.80	−0.20
9	0	45.0	7.89	−1.11
10	0	10.0	8.00	−2.00

$p(A = X) = .50; p(B = X) = .50;$
$p(\pm 1) = .20; p(\pm 2) = .05$ (see text);
$s_x^2 = 2.77; s_A^2 = s_B^2 = 3.57; s_X^2/s_A^2 = 0.78$

observed category $A = 4$ 'moves' .20 toward a higher score because there are more people who are truly above than below 4, but the people in $A = 5$ 'regress' toward the modal category 4, away from the mean. So, the regression phenomenon is not confined to and not only a consequence of the statistical regression model, but may occur in several different ways and directions, and poses a threat not only to the analysis of continuous data, but also to categorical and nominal level data analysis.

2.3 Fickle Minorities: Tendencies Toward the Mode in Transition Tables

The kernel of the analysis of longitudinal categorical data is the turnover table, usually in the form of a transition table, showing the changes over time conditional upon the initial states. Table 2.4 provides such a table in its most elementary form of a 2 × 2 table. From Table 2.4 (Panel A) and its rearrangement in Table 2.4 (Panel B) it may be concluded that those who had a favorable attitude (in this case: children's attitude toward minorities) in the beginning of the study were still overwhelmingly favorable at the second wave, while almost half of those who were unfavorable to

Table 2.4: Cross Tabulation of Opinion About Minorities at Two Points in Time.

	Panel A				Panel B		
	Time 2				Time 2		
	Fav.	Unfav.	Total		Same	Diff.	Total
Time 1				Time 1			
Fav.	89.0	11.0	100	Fav.	89.0	11.0	100
	(81)	(10)	(91)		(81)	(10)	(91)
Unfav.	47.6	52.4	100	Unfav.	52.4	47.6	100
	(10)	(11)	(21)		(11)	(10)	(21)
Total	81.2	18.8	100	Total	82.1	17.9	100
	(91)	(21)	(112)		(91)	(21)	(112)

Note. Horizontal percentages; frequencies between brackets. Source: see text.
Fav. = Favorable; Unfav. = Unfavorable; Diff. = Different.

begin with changed into a favorable attitude. Given that in the meantime a documentary was shown on T.V. providing favorable information about the culture of minority groups, it would have been concluded that the documentary was successful. Or, imagine another research context, in which it was considered to be important that the unfavorable people were a minority in the first wave and that this minority would experience social pressure to share the majority opinion. The empirical results seem to confirm this expectation: the majority was rather stable, but the minority changed into the direction of the majority.

Hundreds, thousands of transition tables can be found that show this pattern: Voters turn out to be much more stable than the smaller group of nonvoters; voters for bigger political parties change much less than supporters of smaller parties; employed people are more stable over time than the unemployed; pupils who choose bigger academic disciplines, such as medicine, follow up that intention much more than those who initially choose for smaller disciplines, such as sociology; owners of a type of car having a big share of the market switch much less to another brand than owners of small brands. There seems to be an omnipresent tendency toward the mode, to the Triumph of Majority in Human Life.

But we know that the study of change is very tricky whenever there may be a random component involved. Let us have a closer look at Table 2.4. Although thousands of exemplary turnover tables might have been provided using real world data, Table 2.4 contains simulated data. So we know how it came about. The patterns of change in Table 2.4 look very much like the classical regression toward the mean. If category favorable is coded '1' and unfavorable '0', the overall mean score at time 1 and time 2 is p, the proportion in category 1, and equals .812 and so the variances $p(1 - p)$ at time 1 and 2 are also equal. The expected (mean) value at time 2 for those who scored '1' the first time is .890. The expected value of the majority group whose score of 1 is close to the overall mean of .81 regressed .078 toward the overall mean. The time 2 expected (mean) value of the minority, with an initial extreme score of 0 far away from the overall mean is .476, regressing .336 toward the overall mean. These regressions are in exact conformity with the correlation (or the identical regression) coefficient in Table 2.3 of .414. However, given the way the data were constructed, the conclusions about the larger stability of the favorables compared to the unfavorables are wrong.

In 'reality', from the total group of 112 children, 101 were assigned to the category truly favorable and 11 to truly unfavorable. It was assumed that these true scores remained completely stable during the period of investigation. The true attitude was measured twice, but not perfectly. All children were given a .90 chance of providing the correct answer, that is, an observed answer in agreement with their true score (yielding of course a probability of misclassification of .10). The probabilities of a misclassification were the same for both points in time and independent of each other. The 'observed' data in Table 2.4 resulted from this simple model. Despite the dramatic observed differences in Table 2.4 regarding the stability of their attitude, the unfavorable children did not truly change more than the favorable ones: nobody's' true position changed and, for example, no effects of a supposed T.V. documentary is present. Moreover, as far as one would like to see the random component (i.e., the probability of a misclassification) as a sign of instability or fickleness, this probability was exactly the same for the majority, the children in favor, and the minority, the children not in favor. In short, all of these kinds of substantive interpretations of the observed changes are completely besides the point given the way the data were simulated.

For the data in Table 2.4, it is sure that the observed changes were an artifact of a small random component. But what if this table had represented real world data. Might we not just take the observed changes for granted, at face value? Because of the omnipresence of random components in our data, the answer is definitely 'no.' First of all, there always will be some random component in our data in the form of measurement error. Even if we look into the measurement of simple variables such as gender, age, education, or religious denomination, we must conclude that these measurements are not or far from perfect (Hagenaars, 1973). How much more likely then must be misclassifications when measuring more complicated concepts, such as values, attitudes, intentions, and beliefs. Furthermore, besides measurement error in the strict sense, there may be other sources of random fluctuations. When stating one's voting preferences or when making a decision which car to buy or which university discipline to study, random components in the form of incidental factors and the mood the people happen to be in will play a role (Kendall, 1954; see also the discussion in the psychometric literature, e.g., Heinen, 1996, chap. 6; Holland, 1990; Lord & Novick, 1968, chap. 2; Lumsden, 1978; Molenaar, 1995). Very small amounts of 'random behavior' and 'misclassifications' may lead to big differences in turnover tables. Defining theoretically sensible, alternative latent variable (latent class) models (such as in Fig. 2.2), estimate (or if necessary 'guess') its parameters and see whether the outcomes make substantively sense may prevent one from confusing regression-like artifacts with substantive changes. Certainly for more complicated situations, such latent variable models may prove necessary to get a good insight into the data.

2.4 Extending the Turnover Table

The discussion in the previous section concentrated on the simple question: how stable is a certain category. But in bigger turnover tables more complex patterns of change might occur. Does the random component also 'determine' into what direction the observed changes will take place? Suppose that in a certain country there is a four party system: A-Labor; B-Conservative; C-Christian-Democratic; D-Extreme Right and that 1000 people truly preferred party A, 400 party B, 400 party C, and only 20 party D. When asked about their party preference the probability of a correct answer is .70 and the probability of mentioning a 'wrong' party is .10 for each of the remaining parties. As usual it is assumed that the true party choice is stable over time and that the (mis)classification probabilities are independent of each other and over time.

Table 2.5 shows what would have been observed for two points in time, given these conditions. First of all, and as expected: The bigger the party, the more loyal its voters seem to be. Particularly the extremely low stability of the smallest party is remarkable. But this recurring pattern toward the mode is also true for the direction of the changes: When the previous party is not chosen again, the change is toward the bigger parties. Especially the initial voters for D change dramatically toward A, much more so than toward B and C, and this despite the fact that in the model the probability of choosing a 'wrong' party is always .10 regardless of the party it concerns. Obvious substantive conclusions such as that the voters for party D are

Table 2.5: Cross Tabulation of Party Choice at Two Points in Time.

		Time 2			
Time 2	A	B	C	D	Total (100=)
A	63.7	13.1	13.1	10.2	*782*
B	24.2	49.8	15.7	10.3	*422*
C	24.2	15.7	49.8	10.3	*422*
D	40.9	22.4	22.4	14.3	*194*
Total	43.0	23.2	23.2	10.7	*1820*

Note. Source see text.

politically much closer to party A than to B or C are clearly false here. Real world tables will most probably show a mixture of real and artificial changes and affinities. One needs appropriate latent class (latent budget) models to disentangle in big turnover tables the two sources of observed change (Clogg 1981; Hagenaars 1990; Van der Heijden, Van der Ark, & Mooijaart, 2002).

Turnover tables are not only studied for the sample as a whole but also to compare the changes among certain subgroups. A simple example of the consequences of misclassifications that may then arise is presented in Table 2.6, again with simulated data. The data are about Vote Intention among old and young people.

We assume that there are 1000 old and 1000 young people. Most of the old people are truly voters (900) and a minority not (100). The young people are evenly divided: 500 true voters, 500 true nonvoters. The probability of a correct answer for both groups is .9. Furthermore, the usual assumptions are made: no true change and independent misclassifications. The only difference between the two groups concerns the distribution of the true scores.

In terms of the amount of simple gross change (a measure in general not to be recommended) old and young people have the same tendency to change: $180/1000 = .18$. Although nobody changed 'in reality,' at least the conclusion that young and old people behave the same is true. However, when looking at the conditional changes or using other 'stability' (association) coefficients, large differences between young and old people are observed. The voters are almost as stable among the old as among the young (.89, .82, respectively), but the nonvoters are very different. Half

Table 2.6: Cross Tabulation of Voting Intentions Among Old (Panel A) and Young (Panel B) at Two Points in Time.

Panel A: Old

	Time 2 Voters	Non-Voters	Total
Time 1			
Voters	89.0	11.0	100
	(730)	(90)	(820)
Non-voters	50.0	50.0	100
	(90)	(90)	(180)
Total	82.0	18.0	100
	(820)	(180)	(1000)

$r = .39$; $\gamma = .78$; odds ratio = 8.11

Panel B: Young

	Time 2 Voters	Non-Voters	Total
Time 1			
Voters	82.0	18.0	100
	(410)	(90)	(500)
Non-voters	18.0	82.0	100
	(90)	(410)	(580)
Total	50.0	50.0	100
	(500)	(500)	(1000)

$r = .64$; $\gamma = .90$; odds ratio = 20.75.

Note. Source see text.

of the old nonvoters voted the second time, compared to about one fifth among the young nonvoters. It looks as if a lot of old nonvoters did not vote for accidental reasons and returned to their usual (voter) self the second time. Young nonvoters on the other hand did not vote out of conviction and remained nonvoter. Such a conclusion, which we know to be false, is corroborated by several association coefficients: Young people are much more stable and show a much larger association between voting at time 1 and voting at time 2 than old people. This is true for the correlation coefficient (.64 vs. .39), for the gamma coefficient (.90 vs. .78) but also for the odds ratio (20.75 vs. 8.11), which was expected to be least influenced by the differing true marginal distributions, but obviously not by misclas-

Table 2.7: Cross Tabulation of Watching T.V. and Opinions About Minorities.

		A (Watching T.V.)		
B (Attitude 1)	C (Attitude 2)	1 (Yes)	2 (No)	Total
1 (Favorable)	1 (Favorable)	81	326	407
	2 (Unfavorable)	10	89	99
2 (Unfavorable)	1 (Favorable)	10	89	99
	2 (Unfavorable)	11	484	495
Total		112	988	1100

sifications.

One of the interesting possibilities of panel studies and subgroup comparisons is to analyze the resulting longitudinal data from a more explicit causal point of view, as coming, for example, from a quasi-experimental design, more in particular from a nonequivalent control group design (Campbell & Stanley, 1966). The simulated data in Table 2.7 is used to illustrate the ('regression') dangers involved in this approach. The data are from an imaginary study into children's attitude toward minority groups (and were partly used in Table 2.4). In this imaginary study, an attempt was made to positively influence the children's attitude by means of a series of T.V. shows about the cultures of the minority groups.

There are several ways of analyzing such data. The two most important ones are the unconditional (or marginal) analysis and the conditional method. In the first approach, one simply subtracts the pretest differences between the 'experimental and the 'control' group from the posttest differences in order to determine the effect of the quasi-experimental factor. In the second approach, one determines the effect of X, the 'experimental' variable, on the posttest, while conditioning on and controlling for the pretest scores. The two approaches generally do not yield the same outcomes and may lead to different conclusions (essentially, Lord's paradox). Only the latter approach will be discussed here, but then from a categorical rather than the usual continuous data point of view (for more general discussions, see Bryk & Weisberg, 1977; Hagenaars, 1990, section 5.3; Judd & Kenny, 1981, chap. 6; Kenny & Cohen, 1979; Lord, 1960, 1967).

The relevant log-linear covariance models for the data in Table 2.7 are presented in Figure 2.3. The saturated model shows a nonzero three-variable interaction but its value is very small (effect coding was used where the effects sum to zero over any subscript) and not statistically significant.

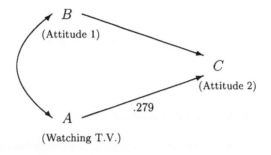

Model $\{ABC\}$: $\qquad$ $\hat{\lambda}_{111}^{ABC} = -.101$ $\quad$ $\sigma_{\hat{\lambda}} = .072$

Model $\{AB, BC\}$: $\qquad$ $L^2 = 17.11$

$\qquad\qquad\qquad\qquad$ $df = 2$ $\qquad$ $p = .00$

Model $\{AB, AC, BC\}$: $\quad$ $L^2 = 1.89$

$\qquad\qquad\qquad\qquad$ $df = 1$ $\qquad$ $p = .1$

$\qquad\qquad\qquad\qquad$ $\hat{\lambda}_{11}^{AC} = .279$ $\quad$ $\sigma_{\hat{\lambda}} = .075$

Figure 2.3: Log-linear Covariance Analysis of the Data in Table 2.7.

Not surprisingly then, the no three variable interaction model $\{AB, AC, BC\}$ fitted the data well and showed a significant, moderate direct effect of A (Watching T.V.) on C (Attitude 2), $(\lambda_{11}^{AC} = .279)$, controlling for the pretest B. Model $\{AB, BC\}$ in which no direct effect of A on C is assumed has to be rejected. The conclusion is clear: The T.V. series was successful (except that not that many children watched). More precisely: The odds of expressing a favorable attitude rather than a unfavorable one at the posttest are three times, $[\exp(.279)]^4 = 3.053$, as high for those who watched the series compared to those that did not watch, controlling for the pretest, that is, within groups that had the same score on the pretest.

Not surprisingly by now, this conclusion is wrong, given the way the data have been constructed. It was assumed that there was a true attitude X that was stable over time. The probability of a misclassification was .10 both at the pretest and at the posttest. The misclassifications were independent of each other. It was further assumed that a favorable attitude made one inclined to watch and that not that many children watched. To be exact: 112 children watched the T.V. Series, 101 (90%) of which were truly favorable and 11 (10%) truly unfavorable; 988 children did not watch, 395 (40%) of which had a true favorable attitude and 593 (60%) not. In this model, nobody's true position changed and watching T.V. did not have any effect on the observed attitude at the time of the posttest. This model generated the data in Table 2.1. It is depicted in Figure 2.4. It can be viewed as a log-linear model with a latent variable X or identically, as a latent class model. The parameter 'estimates' follow directly from the way the data were generated; the λ-parameters denote the parameters of the log-linear model, π_t^X is the probability of belonging to latent (true) class t and $\pi_t^{A|X}$, is the conditional response probability of scoring i on A, given $X = t$.

Given the model that generated the data, the conclusions based on the covariance analysis are obviously wrong. The problem lies in the misclassifications when measuring the attitude. The seemingly clear outcome of the covariance analysis and especially the effect of $\hat{\lambda}_{11}(AC = .279$ is completely the result of the very small amount of misclassifications and the intricate workings of the regression artifact. In general, if the true scores had been known and had also been used as a covariate rather than the fallible pretest scores, no effect would have been found.

It may be clear from the previous discussion that when studying change, small amounts of random error may lead researchers completely astray. Very systematically looking patterns in small and large turnover or mobility tables, in comparisons of subgroups, and in data from the nonequivalent control group design may just be appearances of regression artifacts and tendencies toward the mode. Other kinds of panel analyses not mentioned here may suffer from the same problems. The phenomenon that people that changed in the past are more likely to change their position again in the next wave than people who were stable in the past may also arise from a situation where nobody's true position changed in combination with small observation errors. Or take Paul Lazarsfeld's (1972c) ingenious use of the observed patterns in the sixteenfold table resulting from two dichotomous characteristics, for example, party preference (Democratic/Republican) and presidential candidate preference (Democrat/Republican) measured at two points in time. Lazarsfeld uses such a table to determine which characteristic causes which from the way inconsistent positions at time one (preference for the Democratic party but the Republican candidate, and so on) are resolved at time 2. But also these patterns may result from sheer measurement error and most of the tables he presents can also be explained in terms of two characteristics that are stable over time without the possibility to infer from the outcomes the causal order of the variables (Goodman, 1974b; Hagenaars, 1990; Lazarsfeld, 1972a; see also the literature on the cross-lagged panel correlation technique, mentioned in Hagenaars, 1990, pp. 240-241). Luckily, the means to prevent this confusion of true change and change due to random components is available. Latent variables can be

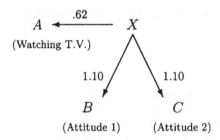

Effects: Estimated λ-Parameters (Effect Coding).

Model $\{AX, BX, CX\}$:

$$\ln \pi_{ijkt}^{ABCX} = \lambda + \lambda_i^A + \lambda_j^B + \lambda_k^C + \lambda_t^X + \lambda_{it}^{AX} + \lambda_{jt}^{BX} + \lambda_{kt}^{CX}$$

$$\pi_{ijkt}^{ABCX} = \pi_t^X \, \pi_{it}^{A|X} \, \pi_{jt}^{B|X} \, \pi_{kt}^{C|X}$$

	$\hat{\pi}_t^X$	$\hat{\pi}_{it}^{A|X}$		$\hat{\pi}_{jt}^{B|X}$		$\hat{\pi}_{kt}^{C|X}$	
		$i = 1$	$i = 2$	$j = 1$	$j = 2$	$k = 1$	$k = 2$
$t = 1$	.45	.20	.80	.90	.10	.90	.10
$t = 2$	.55	.02	.98	.10	.90	.10	.90

Figure 2.4: Latent Class Analysis of the Data in Table 2.7.

introduced in covariance analyses; marginal tests on the equality of means or odds at the latent level have been introduced recently (Bergsma, Croon, & Hagenaars, 2004; Hagenaars 1998). Researchers should take these latent models seriously as an alternative explanation of what seemingly has been observed. Above, artificial data have been used to make the argument more convincing: We knew what was going on. But hundreds of real world examples might be given showing the same kinds of patterns and results. For example, an extended real world example about unemployment status is provided by Bassi et al. (2000), in which misclassifications occur in combination with true latent changes, and in which the misclassifications are no longer independent, but may show rather complex dependencies.

2.5 Discussion

The frequent occurrence of regression toward the mean, or rather the tendency toward the mode, is essentially due to the omnipresence of random fluctuations, whether or not in combination with correlated 'error.' But what is actually understood by error, misclassifications, and random components or equivalently, what are true scores and latent variables? First, a distinction must be made between latent variables and true scores. Whereas true scores always refer to latent scores, the opposite is not true. For example, in factor analysis or latent class analysis, latent variables are often used as a kind of summary measures, replacing a much bigger amount of observed variables or categories without the notion that these latent variables refer to true properties a person possesses.

Lord and Novick (1968, pp. 27-54) have presented an in-depth analysis of the meanings of true scores (see also Suttcliffe 1965a, 1965b). Essentially they defend the view that the true score is the expected value coming out of a series of independent experiments, that is, out of independent replications of the same measurement instrument. Another way to look at the true score is as a platonic score: A person has a true score, which in principle exists and can be known, but is measured with error. Weight or Marital Status might be considered as true platonic scores, while true attitude scores are more in line with the 'expectation' definition.

There are several reasons for error, for not observing directly the true score. There is of course proper measurement error, interviewer mistakes, and so on.

> But why adopt this idea [of proper measurement error—JH] from a psychometric tradition? Certainly if a man changes his vote intention it does not mean that the interview is ambiguous; much more likely a multiplicity of influences make him oscillate

around his "true" position from one interview to the next. And
in the case of mood [...] every reader knows only too well how
realistic change is. Mood shifts would manifest themselves irre-
spective of the way they are "measured" (Lazarsfeld, 1972b, p.
360).

Lazarsfeld essentially distinguishes three sources of observed instability:
changes in true scores and changes due to two random components, namely
'real' random fluctuations around true scores caused by incidental factors,
and proper measurement errors. Latent variable models 'correct' the data
for both types of errors, whether due to 'real' random fluctuations or to
random errors of measurement (ignoring here the possibilities of taking
into account correlated errors and systematic misclassifications). For many
purposes, this throwing together of the two random components does not
matter. But sometimes it does. If one visits the doctor and this doctor
wants to know your weight, she or he is usually not really interested in the
exact 'platonic' weight at that precise moment, having had that amount
of coffee, two biscuits for breakfast, and so on. But she or he is interested
in the average (expected) weight over a certain period, which may be obtained by weighing a couple of times, controlling for the incidental factors
and the measurement errors when reading off the balance. But maybe, just
before a complicated surgical operation, one's exact weight at that particular moment may be of interest. Or in terms of our labor force status
example, the estimates obtained from a latent class cannot always be used
as estimates for the true number of people that are actually unemployed
at a certain moment. The latent class model does not distinguish between
random fluctuations due to proper measurement error and 'real' random
fluctuations due to incidental factors. That distinction can only be made if
a perfect measurement, a is available. Whether that is necessary depends
on the exact purposes of the investigation. In any case, latent variable
models provide a powerful means of recognizing or even beating the misleading effects of the regression phenomenon. But, remembering Stigler's
words from the beginning: Will they be used and will we ever learn?

References

Bassi, F., Hagenaars, J. A., Croon, M., & Vermunt, J. K. (2000). Estimating true changes when categorical panel data are affected by un-
correlated and correlated errors. *Sociological Methods and Research,*
29, 230-268.

Bergsma, W. P., Croon, M., & Hagenaars, J. A. (2004). *Marginal models*
for categorical data. Manuscript in preparation.

Bryk, A. S., & Weisberg, H. I. (1977). Use of the nonequivalent control group design when subjects are growing. *Psychological Bulletin, 84*, 950-962.

Campbell, D. T., & Clayton, K. N. (1961). Avoiding regression effects in panel studies of communication impact. *Studies in Public Communication, 3*, 99-118. Chicago: University of Chicago, Department of Sociology (Bobbs-Merrill Reprint Series No. S 353).

Campbell, D. T., & Kenny, D. A. (1999). *A primer on regression artifacts*. New York: Guilford Press.

Campbell, D. T., & Stanley, J. C. (1966). *Experimental and quasi-experimental designs for research*. Chicago: Rand McNally.

Clogg, C. C. (1981). Latent structure models of mobility. *American Journal of Sociology, 86*, 836-868.

Desrosières, A. (1998). *The politics of large numbers; A history of statistical reasoning*. Cambridge, MA: Harvard University Press.

Friedman, M. (1992). Do old fallacies ever die? *Journal of Economic Literature, 30*, 2129-2132.

Goodman, L. A. (1974a). Exploratory latent structure analysis using both identifiable and unidentifiable models. *Biometrika, 61*, 215-231.

Goodman, L. A. (1974b). The analysis of systems of qualitative variables when some of the variables are unobservable. Part I - a modified latent structure approach. *American Journal of Sociology, 79*, 1179-1259.

Haberman, S. J. (1974). Loglinear models for frequency tables derived by indirect observation. Maximum likelihood equations. *The Annals of Statistics 2*, 911-924.

Hagenaars, J. A. (1973). Betrouwbaarheid van meting van achtergrondvariabelen d.m.v. interviews; omvang, oorzaken en konsekwenties [Reliability of the measurement of background variables using interviews; magnitude, causes, and consequences.] *Sociale Wetenschappen, 16*, 236-273.

Hagenaars, J. A. (1975). Problemen bij de analyse van panel gegevens [Problems in the analysis of panel data.] *Mens en Maatschappij, 50*, 73-88.

Hagenaars, J. A. (1990). *Categorical longitudinal data; Log-linear panel, trend, and cohort analysis*. Newbury Park, CA: Sage.

Hagenaars, J. A. (1993). *Loglinear models with latent variables*. Sage University Papers. Newbury Park, CA: Sage.

Hagenaars, J. A. (1994). Latent variables in log-linear models of repeated observations. In A. von Eye & C. C. Clogg (Eds.), *Latent variables analysis; applications for developmental research* (pp. 329-352). Thousand Oaks, CA: Sage.

Hagenaars J. A. (1998). Categorical causal modeling; latent class analysis and directed log-linear models with latent variables. *Sociological Methods and Research, 26*, 436-486.

Hagenaars, J. A., & McCutcheon, A. L. (Eds.). (2002). *Applied latent class analysis.* New York: Cambridge University Press.

Harper, D. (1973). Observation errors in sociological surveys: a model and a method. *Sociological Methods and Research, 2*, 63-83.

Heinen, T. (1996). *Latent class and discrete latent trait models: Similarities and differences.* Thousand Oaks, CA: Sage.

Holland, P. W. (1990). On the sampling theory foundations of item response theory models. *Psychometrika, 55*, 5-18.

Judd, C. M., & Kenny, D. A. (1981). *Estimating the effects of social interventions.* Cambridge: Cambridge University Press.

Kendall, P. (1954). *Conflict and mood; factors affecting stability of responses.* Glencoe, IL: Free Press.

Kenny, D. A., & Cohen, S. H. (1979). Reexamination of selection and growth processes. In K. F. Schuessler (Ed.), *Sociological Methodology 1980* (pp. 290-313). San Francisco: Jossey Bass.

Kuha, J., & Skinner, C. (1997). Categorical data and misclassification. In L. Lyberg, P. Biemer, M. Collins, E. de Leeuw, C. Dippo, N. Schwartz, & D. Trewin, *Survey measurement and process quality* (pp. 633-671). New York: Wiley

Langeheine, R., & Van de Pol, F. (2002). Latent Markov chains. In J. A. Hagenaars & A. L. McCutcheon (Eds.), *Applied latent class analysis* (pp. 304-344). New York: Cambridge University Press.

Lazarsfeld, P. F. (1950). The logical and mathematical foundation of latent structure analysis. In S. A. Stouffer, L. Guttman, E. A. Suchman, P. F. Lazarsfeld, S. A. Star & J. A. Clausen (Eds.), *Studies in social psychology in World War II; Vol. 4. Measurement and prediction* (pp. 362-472). Princeton, NJ: Princeton University Press.

Lazarsfeld, P. F., & Henry, N. W. (1969). *Latent structure analysis.* Boston: Houghton Mifflin.

Lazarsfeld, P. F. (1972a). The use of panels in social research. In P. F. Lazarsfeld, A. K. Pasanella & M. Rosenberg (Eds.), *Continuities in the language of social research* (pp. 330-337). New York: Free Press.

Lazarsfeld, P. F. (1972b). The problem of measuring turnover. In P. F. Lazarsfeld, A. K. Pasanella & M. Rosenberg (Eds.), *Continuities in the language of social research* (pp. 358-367). New York: Free Press.

Lazarsfeld, P. F. (1972c). Mutual effects of statistical variables. In P. F. Lazarsfeld, A. K. Pasanella & M. Rosenberg (Eds.), *Continuities in the language of social research* (pp. 388-398). New York: Free Press.

Lumsden, J. (1978). Tests are perfectly reliable. *British Journal of Mathematical and Statistical Psychology, 31*, 19-26.

Lord, F. M. (1960). Large-sample covariance analysis when the control variable is fallible. *Journal of the American Statistical Association, 55*, 309-321.

Lord, F. M. (1967). A paradox in the interpretation of group comparisons. *Psychological Bulletin, 68*, 304-305.

Lord, F. M., & Novick, M. R. (1968). *Statistical theories of mental test scores.* Reading, MA: Addison-Wesley.

Maccoby, E. E. (1956). Pitfalls in the analysis of panel data; a research note on some technical aspects of voting. *American Journal of Sociology, 61*, 359-362.

McNemar, Q. (1940). A critical examination of the Iowa Studies of environmental influence upon IQ. *Psychological Bulletin, 37*, 63-92.

Molenaar I. W. (1995). Some background for item response theory and the Rasch model. In G. H. Fischer & I. W. Molenaar (Eds.), *Rasch models; Foundations, recent developments, and applications.* New York: Springer.

Poulsen, C. S. (1982). *Latent structure analysis with choice modeling applications.* Unpublished doctoral dissertation, The Århus School of Business Administration and Economics, Denmark.

Reichardt, C. S. (1999). Foreword. In D. T. Campbell, & D. A. Kenny, *A primer on regression artifacts.* New York: Guilford Press.

Stigler, S. M. (1997). Regression toward the mean, historically considered. *Statistical Methods in Medical Research, 6*, 103-114.

Stigler, S. M. (1999). *Statistics on the table; the history of statistical concepts and methods.* Cambridge, MA: Harvard University Press

Suttcliffe, J. P. (1965a). A probability model for errors of classification; I: General considerations. *Psychometrika, 30*, 73-96.

Suttcliffe, J. P. (1965b). A probability model for errors of classification; II: Particular Cases. *Psychometrika, 30*, 129-155.

Thorndike, R. L. (1942). Regression fallacies in the matched group experiment. *Psychometrika, 7*, 85-102.

Van de Pol, F., & Langeheine, R. (1997). Separating change and measurement error in panel surveys with an application to labor market data. In L. Lyberg, P. Biemer, M. Collins, E. de Leeuw, C. Dippo, N. Schwartz & D. Trewin (Eds.), *Survey measurement and process quality* (pp. 671-688). New York: Wiley.

Van der Heijden, P. G. M., Van der Ark, L. A., & Mooijaart, A. (2002). Some examples of latent budget analysis and its extensions. In J. A.

Hagenaars & A. L. McCutcheon (Eds.), *Applied latent class analysis* (pp. 107-136). New York: Cambridge University Press.

Vermunt, J. K. (1997). *Loglinear models for event history analysis.* Thousand Oaks, CA: Sage.

Wiggins, L. M. (1955). *Mathematical models for the analysis of multiwave panels* (Doctoral Dissertation Series No. 12.481). Ann Arbor MI: University of Michigan.

Chapter 3

Factor Analysis With Categorical Indicators: A Comparison Between Traditional and Latent Class Approaches

Jeroen K. Vermunt
Tilburg University

Jay Magidson
Statistical Innovations, Inc.

3.1 Introduction

The linear factor analysis (FA) model is a popular tool for exploratory data analysis or, more precisely, for assessing the dimensionality of sets of items. Although it is well known that it is meant for continuous observed indicators, it is often used with dichotomous, ordinal, and other types of discrete variables, yielding results that might be incorrect. Not only parameter estimates may be biased, but also goodness-of-fit indices cannot be trusted. Magidson and Vermunt (2001) presented a nonlinear factor-analytic model

Table 3.1: Four-Fold Classification of Latent Variable Models.

Manifest variables	Latent variables	
	Continuous	Categorical
Continuous	Factor analysis	Latent profile analysis
Categorical	Latent trait analysis	Latent class analysis

based on latent class (LC) analysis that is especially suited for dealing with categorical indicators, such as dichotomous, ordinal, and nominal variables, and counts. The approach is called latent class factor analysis (LCFA) because it combines elements from LC and traditional FA. This LCFA model is one of the LC models implemented in the Latent GOLD program (Vermunt & Magidson, 2000, 2003).

A disadvantage of the LCFA model is, however, that its parameters may be somewhat more difficult to interpret than the typical factor-analytic coefficients—factor loadings, factor-item correlations, factor correlations, and communalities. To overcome this problem, we propose using a linear approximation of the maximum likelihood estimates obtained with a LCFA model. This makes it possible to provide the same type of output measures as in standard FA, while retaining the fact that the underlying factor structure is identified by the more reliable nonlinear factor-analytic model.

Bartholomew and Knott (1999) gave a four-fold classification of latent variable models based on the scale types of the latent and observed variables; that is, factor analysis, latent trait (LT) analysis, latent profile (LP) analysis, and latent class analysis. As shown in Table 3.1, in FA and LT models, the latent variables are treated as continuous normally distributed variables. In LP and LC models on the other hand, the latent variable is assumed to be discrete and to come from a multinomial distribution. The manifest variables in FA and LP model are continuous. In most cases, their conditional distribution given the latent variables is assumed to be normal. In LT and LC analysis, the indicators are dichotomous, ordinal, or nominal categorical variables and their conditional distributions are assumed to be binomial or multinomial.

The distinction between models for continuous and discrete indicators is not a fundamental one since the choice between the two should simply depend on the type of data. The specification of the conditional distributions of the indicators follows naturally from their scale types. A recent development in latent variable modelling is to allow for a different distributional form for each indicator. This can, for example, be a normal, student, log-normal, gamma, or exponential distribution for continuous variables, binomial for dichotomous variables, multinomial for ordinal and nominal variables, and Poisson, binomial, or negative-binomial for counts. Depending on whether the latent variable is treated as continuous or discrete, one obtains a generalized LT (Moustaki & Knott, 2000) or LC (Vermunt & Magidson, 2001) model.

The more fundamental distinction in Bartholomew's typology is the one between continuous and discrete latent variables. A researcher has to decide whether to treat the underlying latent variable(s) as continuous or discrete. However, Heinen (1996) demonstrated that the distribution of a continuous latent variable can be approximated by a discrete distribution, and that such a discrete approximation may even be superior[1] to a misspecified continuous (usually normal) model. More precisely, Heinen (1996; also, see Vermunt, 2001) showed that constrained LC models can be used to approximate well-known unidimensional LT or item response theory (IRT) models, [2] such as the Rasch, the Birnbaum, the nominal-response, and the partial credit model. This suggests that the distinction between continuous and discrete latent variables is less fundamental than one might initially think, especially if the number of latent classes is increased. More precisely, as shown by Aitkin (1999; also, see Vermunt and Van Dijk, 2001; Vermunt, 2004), a continuous latent distribution can be approximated using a nonparametric specification; that is, by a finite mixture model with the maximum number of identifiable latent classes. An advantage of such a nonparametric approach is that it is not necessary to introduce possibly inappropriate and unverifiable assumptions about the distribution of the random effects.

The proposed LCFA model is based on a multidimensional generalization of Heinen's (1996) idea: It is a restricted LC model with several latent variables. As exploratory FA, the LCFA model can be used to determine which items measure the same dimension. The idea of defining an LC model with several latent variables is not new: Goodman (1974) and Hagenaars (1990) proposed such a model and showed that it can be derived from a standard LC model by specifying a set of equality constraints on the item conditional probabilities. What is new is that we use IRT-like regression-type constraints on the item conditional means/probabilities[3] to be able

---

[1]With superior we refer to the fact that misspecification of the distribution of the continuous latent variables may cause bias in the item parameter estimates. In a discrete or nonparametric specification,, no assumptions are made about the latent distribution, and as a result, parameters cannot be biased because of misspecification of the latent distribution.

[2]We use the terms latent trait (LT) and item response theory (IRT) interchangeably.

[3]With regression-type constraints on the item conditional probabilities we mean that

to use the LC model with several latent variables as an exploratory factor-analytic tool. Our approach is also somewhat more general than Heinen's in the sense that it cannot only deal with dichotomous, ordinal, and nominal observed variables, but also with counts and continuous indicators, as well as any combination of these.

Using a general latent variable model as the starting point, it will be shown that several important special cases are obtained by varying the model assumptions. In particular, assuming that the latent variables are dichotomous or ordinal, and that the effects of these latent variables on the transformed means are additive, yields the proposed LCFA model. We show how the results of this LCFA model can be approximated using a linear FA model, which yields the well-known standard FA output. Special attention is given to the meaning of the part that is ignored by the linear approximation and to the handling of nominal variables. Several real life examples are presented to illustrate our approach.

3.2 The Latent Class Factor Model

Let $\boldsymbol{\theta}$ denote a vector of L latent variables and $\mathbf{y}$ a vector of K observed variables. Indices ℓ and k are used when referring to a specific latent and observed variable, respectively. A basic latent variable model has the following form

$$f(\boldsymbol{\theta}, \mathbf{y}) = f(\boldsymbol{\theta}) f(\mathbf{y}|\boldsymbol{\theta}) = f(\boldsymbol{\theta}) \prod_{k=1}^{K} f(y_k|\boldsymbol{\theta}),$$

where the primary model assumption is that the K observed variables are independent of one another given the latent variables $\boldsymbol{\theta}$, usually referred to as the local independence assumption (Bartholomew and Knott, 1999). The various types of latent variable models are obtained by specifying the distribution of the latent variables $f(\boldsymbol{\theta})$ and the K conditional item distributions $f(y_k|\boldsymbol{\theta})$. The two most popular choices for $f(\boldsymbol{\theta})$ are continuous multivariate normal and discrete nominal. The specification for the error functions $f(y_k|\boldsymbol{\theta})$ will depend on the scale type of indicator k.[4] Besides the distributional form of $f(y_k|\boldsymbol{\theta})$, an appropriate link or transformation function $g(\cdot)$ is defined for the expectation of y_k given $\boldsymbol{\theta}$, $E(y_k|\boldsymbol{\theta})$. With continuous $\boldsymbol{\theta}$ (FA or LT), the effects of the latent variables are assumed to

---

the probability of giving a particular response given the latent traits is restricted by means of a logistic regression model, or another type of regression model. In the case of continuous responses, the means are restricted by linear regression models, as in standard factor analysis.

[4]The term error function is jargon from the generalized linear modelling framework. Here it refers to the distribution of the unexplained or unique part (the error) of y_k.

be additive in $g(\cdot)$; that is

$$g[E(y_k|\boldsymbol{\theta})] = \beta_{0k} + \sum_{\ell=1}^{L} \beta_{\ell k}\theta_{\ell}, \qquad (3.1)$$

where the regression intercepts β_{0k} can be interpreted as "difficulty" parameters and the slopes $\beta_{\ell k}$ as "discrimination" parameters. With a discrete $\boldsymbol{\theta}$ (LC or LP), usually no constraints are imposed on $g[E(y_k|\boldsymbol{\theta})]$.

The new element of the LCFA model is that a set of discrete latent variables is explicitly treated as multidimensional, and that the same additivity of their effects is assumed as in Equation 3.1. In the simplest specification, the latent variables are specified to be dichotomous and mutually independent, yielding what we call the *basic LCFA model*. An LCFA model with L dichotomous latent variables is, actually, a restricted LC model with 2^L latent classes (Magidson & Vermunt, 2001). Our approach is an extension of Heinen's work to the multidimensional case. Heinen (1996) showed that LC models with certain log-linear constraints yield discretized versions of unidimensional LT models. The proposed LCFA model is a discretized multidimensional LT or IRT model. With dichotomous observed variables, for instance, we obtain a discretized version of the multidimensional two-parameter logistic model (Reckase, 1997).

A disadvantage of the (standard) LC model compared to the LT and LCFA models is that it does not explicitly distinguish different dimensions, which makes it less suited for dimensionality detection. Disadvantages of the LT model compared to the other two models are that it makes stronger assumptions about the latent distribution and that its estimation is computationally much more intensive, especially with more than a few dimensions. Estimation of LT models via maximum likelihood requires numerical integration: for example, with 3 dimensions and 10 quadrature points per dimension, computation of the log-likelihood function involves summation over 1000 ($=10^3$) quadrature points. The LCFA model shares the advantages of the LT model, but is much easier to estimate, which is a very important feature if one wishes to use the method for exploratory purposes. Note that a LCFA model with three dimensions requires summation over no more than eight ($=2^3$) discrete nodes. Of course, the number of nodes becomes larger with more than two categories per latent dimension, but will still be much smaller than in the corresponding LT model.

Let us first consider the situation in which all indicators are dichotomous. In that case, the most natural choices for $f(y_k|\boldsymbol{\theta})$ and $g(\cdot)$ are a binomial distribution function and a logistic transformation function. Alternatives to the logistic transformation are probit, log-log, and complementary log-log transformations. Depending on the specification of $f(\boldsymbol{\theta})$ and model for $g[E(y_k|\boldsymbol{\theta})]$, we obtain a LT, LC, or LCFA model. In the LCFA model

$$f(\boldsymbol{\theta}) \quad = \quad \pi(\boldsymbol{\theta}) = \prod_{\ell=1}^{L} \pi(\theta_\ell)$$

$$g[E(y_k|\boldsymbol{\theta})] \quad = \quad \log\left[\frac{\pi(y_k|\boldsymbol{\theta})}{1 - \pi(y_k|\boldsymbol{\theta})}\right] = \beta_{0k} + \sum_{\ell=1}^{L} \beta_{\ell k}\theta_\ell. \qquad (3.2)$$

The parameters to be estimated are the probabilities $\pi(\theta_\ell)$ and the coefficients β_{0k} and $\beta_{\ell k}$. The number of categories of each of the L discrete latent variables is at least two, and θ_ℓ are the fixed category scores assumed to be equally spaced between 0 and 1. The assumption of mutual independence between the latent variables θ_ℓ can be relaxed by incorporation two-variable associations in the model for $\pi(\boldsymbol{\theta})$. Furthermore, the number of categories of the factors can be specified to be larger than two: A two-level factor has category scores 0 and 1 for the factor levels, a three-level factor scores 0, 0.5, and 1, and so forth.

The above LCFA model for dichotomous indicators can easily be extended to other types of indicators. For indicators of other scale types, other distributional assumption are made and other link functions are used. Some of the possibilities are described in Table 3.2. For example, the restricted logit model we use for ordinal variables is an adjacent-category logit model. Letting s denote one of the S_k categories of variable y_k, it can be defined as

$$\log\left[\frac{\pi(y_k = s|\boldsymbol{\theta})}{\pi(y_k = s - 1|\boldsymbol{\theta})}\right] = \beta_{0ks} + \sum_{\ell=1}^{L} \beta_{\ell k}\theta_\ell,$$

for $2 \leq s \leq S_k$.

Extensions of the basic LCFA model are among others that local dependencies can be included between indicators and that covariates may influence the latent variables and the indicators (Magidson & Vermunt, 2001, 2004). These are similar to extensions proposed for the standard latent class model (e.g., see Dayton & Macready, 1988; Hagenaars, 1988; Van der Heijden, Dessens & Böckenholt, 1996).

Similarly to standard LC models and IRT models, the parameters of a LCFA model can be estimated by means of maximum likelihood (ML). We solved this ML estimation problem by means of a combination of an EM and a Newton-Raphson algorithm. More specifically, we start with EM and switch to Newton-Raphson when close to the maximum likelihood solution. The interested reader is referred to Vermunt and Magidson (2000, Appendix).

Table 3.2: Distribution and Transformation Functions From Generalized Linear Modelling Family.

Scale type y_k	Distribution $f(y_k|\boldsymbol{\theta})$	Transformation $g\left[E(y_k|\boldsymbol{\theta})\right]$
Dichotomous	Binomial	Logit
Nominal	Multinomial	Logit
Ordinal	Multinomial	Restricted logit
Count	Poisson	Log
Continuous	Normal	Identity

3.3 Linear Approximation

As just mentioned, the proposed nonlinear LCFA model is estimated by means of ML. However, as a result of the scale transformations $g(\cdot)$, the parameters of the LCFA model are more difficult to interpret than the parameters of the traditional FA model. To facilitate the interpretation of the results, we propose approximating the maximum likelihood solution for the conditional means $\widehat{E}(y_k|\boldsymbol{\theta})$ by a linear model, yielding the same type of output as in traditional FA. Although the original model for item k may, for example, be a logistic model, we approximate the logistic response function by means of a linear function.

The ML estimates $\widehat{E}(y_k|\boldsymbol{\theta})$ are approximated by the following linear function

$$\widehat{E}(y_k|\boldsymbol{\theta}) = b_{0k} + \sum_{\ell=1}^{L} b_{\ell k}\theta_\ell + e_{k|\boldsymbol{\theta}}. \tag{3.3}$$

The parameters of the K linear regression models are simply estimated by means of ordinary least squares (OLS). The residual term $e_{k|\boldsymbol{\theta}}$ is needed because the linear approximation will generally not be perfect.

With two dichotomous factors, a perfect description by a linear model is obtained by

$$\widehat{E}(y_k|\theta_1, \theta_2) = b_{k0} + b_{k1}\theta_1 + b_{k2}\theta_2 + b_{k12}\theta_1\theta_2;$$

that is, by the inclusion of the interaction between the two factors. Because the similarity with standard FA would otherwise get lost, interaction terms such as b_{k12} are omitted in the approximation.

Special provisions have to be made for ordinal and nominal variables. Because of the adjacent-category logit model specification for ordinal variables, it is most natural to define $\widehat{E}(y_k|\boldsymbol{\theta}) = \sum_{s=1}^{S} s\,\widehat{\pi}(y_k = s|\boldsymbol{\theta})$.[5] With nominal variables, analogous to the Goodman and Kruskal tau-b (GK-τ_b), each category is treated as a separate dichotomous variable, yielding one coefficient per category. These category-specific coefficients are combined into a single measure in exactly the same way as is done in the computation of the GK-τ_b coefficient. As is shown next, overall measures for nominal variables are defined as weighted averages of the category-specific coefficients.

The coefficients reported in traditional linear FA are factor loadings $(p_{\theta_\ell y_k})$, factor correlations $(r_{\theta_\ell \theta_{\ell'}})$, communalities or proportion explained item variances $(R_{y_k}^2)$, factor-item correlations $(r_{\theta_\ell y_k})$, and in the case that there are local dependencies, also residual item correlations $(r_{e_k e_{k'}})$. The correlations $r_{\theta_\ell \theta_{\ell'}}$, $r_{\theta_\ell y_k}$, and $r_{y_k y_k}$ can be computed from $\widehat{\pi}(\boldsymbol{\theta})$, $\widehat{E}(y_k|\boldsymbol{\theta})$, and the observed item distributions using elementary statistics computation. For example, the $r_{\theta_\ell \theta_{\ell'}}$ is obtained by dividing the covariance between θ_ℓ and $\theta_{\ell'}$ by the product of their standard deviations; that is

$$r_{\theta_\ell \theta_{\ell'}} = \frac{\sigma_{\theta_\ell \theta_{\ell'}}}{\sigma_{\theta_{\ell'}} \sigma_{\theta_{\ell'}}} = \frac{\sum_{\theta_\ell} \sum_{\theta_{\ell'}} [\theta_\ell - \widehat{E}(\theta_\ell)][\theta_{\ell'} - \widehat{E}(\theta_{\ell'})]\,\widehat{\pi}(\theta_\ell \theta_{\ell'})}{\sqrt{\sum_{\theta_\ell} [\theta_\ell - \widehat{E}(\theta_\ell)]^2\,\widehat{\pi}(\theta_\ell)} \sqrt{\sum_{\theta_{\ell'}} [\theta_{\ell'} - \widehat{E}(\theta_{\ell'})]^2\,\widehat{\pi}(\theta_{\ell'})}},$$

where $\widehat{E}(\theta_\ell) = \sum_{\theta_\ell} \theta_\ell\,\widehat{\pi}(\theta_\ell)$.

The factor-factor $(r_{\theta_\ell \theta_{\ell'}})$ and the factor-item $(r_{\theta_\ell y_k})$ correlations can be used to compute OLS estimates for the factor loadings $(p_{\theta_\ell y_k})$, which are standardized versions of the regression coefficients appearing in Equation 3.3. The communalities or R^2 values $(R_{y_k}^2)$ corresponding to the linear approximation are obtained with $r_{\theta_\ell y_k}$ and $p_{\theta_\ell y_k}$: $R_{y_k}^2 = \sum_{\ell=1}^{L} r_{\theta_\ell y_k}\, p_{\theta_\ell y_k}$. The residual correlations $(r_{e_k e_{k'}})$ are defined as the difference between $r_{y_k y_{k'}}$ and the total correlation (not only the linear part) induced by the factors, denoted by $r_{y_k y_{k'}}^{\boldsymbol{\theta}}$.

The linear approximation of $\widehat{E}(y_k|\boldsymbol{\theta})$ is, of course, not perfect. One error source is caused by the fact that the approximation excludes higher-order interaction effects of the factors. More specifically, in the LCFA model presented in Equation 3.2, higher-order interactions are excluded, but this does not mean that no higher-order interactions are needed to get a perfect linear approximation. Alternatively, with all interaction included, the linear approximation would be perfect. For factors having more than two ordered levels, there is an additional error source caused by the fact that linear effects on the transformed scale are nonlinear on the untransformed scale.

---

[5] The same would apply with other link functions for ordinal variables, such as with a cumulative logit link.

To get insight in the quality of the linear approximation, we also compute the R^2 treating the joint latent variable as a set of dummies; that is, as a single nominal latent variable.

As was mentioned above, for nominal variables, we have a separate set of coefficients for each of the S_k categories because each category is treated as a separate dichotomous indicator. If s denotes one of the S_k categories of y_k, the category-specific R^2 ($R^2_{y_k^s}$) equals

$$R^2_{y_k^s} = \frac{\sigma^2_{\hat{E}(y_k=s|\boldsymbol{\theta})}}{\sigma^2_{y_k^s}},$$

where $\sigma^2_{\hat{E}(y_k=s|\boldsymbol{\theta})}$ is the explained variance of the dummy variable corresponding to category s of item k, and $\sigma^2_{y_k^s}$ is its total variance defined as $\pi(y_k=s)[1-\pi(y_k=s)]$. The overall $R^2_{y_k}$ for item k is obtained as a weighted sum of the S_k category-specific R^2 values, where the weights $w_{y_k^s}$ are proportional to the total variances $\sigma^2_{y_k^s}$; that is

$$R^2_{y_k} = \sum_s^{S_k} \frac{\sigma^2_{y_k^s}}{\sum_t^{S_k} \sigma^2_{y_k^t}} R^2_{y_k^s} = \sum_s^{S_k} w_{y_k^s} R^2_{y_k^s}.$$

This weighting method is equivalent to what is done in the computation of the GK-τ_b, an asymmetric association measure for nominal dependent variables.

We propose using the same weighting in the computation of $p_{\theta_\ell y_k}$, $r_{\theta_\ell y_k}$, and $r_{e_k e_{k'}}$ from their category-specific counterpart. This yields

$$p_{\theta_\ell y_k} = \sqrt{\sum_{s=1}^{S_k} w_{y_k^s} \left(p_{\theta_\ell y_k^s}\right)^2}$$

$$r_{\theta_\ell y_k} = \sqrt{\sum_{s=1}^{S_k} w_{y_k^s} \left(r_{\theta_\ell y_k^s}\right)^2}$$

$$r_{e_k e_{k'}} = \sqrt{\sum_{s=1}^{S_k} \sum_{t=1}^{S_{k'}} w_{y_k^s} w_{y_{k'}^t} \left(r_{e_k^s, e_{k'}^t}\right)^2}.$$

As can be seen the signs are lost, but that is, of course, not a problem for a nominal variable.

3.4   Empirical Examples

3.4.1   Rater Agreement

For our first example we factor analyze dichotomous ratings made by seven pathologists, each of whom classified 118 slides as to the presence or absence of carcinoma in the uterine cervix (Landis & Koch, 1977). This is an example of an inter-rater agreement analysis. We want to know whether the ratings of the seven raters are similar, and if not, in what sense the ratings deviate from each other.

Agresti (2002), using standard LC models to analyze these data found that a two-class solution does not provide an adequate fit to these data. Using the LCFA framework, Magidson and Vermunt (2004) confirmed that a single dichotomous factor (equivalent to a two-class LC model) did not fit the data. They found that a basic two-factor LCFA model provides a good fit.

Table 3.3 presents the results of the two-factor model in terms of the conditional probabilities. These results suggest that Factor 1 distinguishes between slides that are "true negative" or "true positive" for cancer. The first class ($\theta_1 = 0$) is the "true negative" group because it has lower probabilities of a "+" rating for each of the raters than class two ($\theta_1 = 1$), the "true positive" group. Factor 2 is a bias factor, which suggests that some pathologists bias their ratings in the direction of a "false +" error ($\theta_2 = 1$) and others exhibit a bias toward "false –" error ($\theta_2 = 0$). More precisely, for some raters we see a too high probability of a "+" rating if $\theta_1 = 0$ and $\theta_2 = 1$ (raters A, G, E, and B) and for others we see a too high probability of a "–" rating if $\theta_1 = 1$ and $\theta_2 = 0$ (raters F and D). These results demonstrate the richness of the LCFA model to extract meaningful information from these data. Valuable information includes an indication of which slides are positive for carcinoma,[6] as well as estimates of "false +" and "false –" error for each rater.

The left-most columns of Table 3.4 lists the estimates of the logit coefficients for these data. Although the probability estimates in Table 3.3 are derived from these quantities (recall Equation 3.2), the logit parameters are not as easy to interpret as the probabilities. For example, the logit effect of θ_1 on A, a measure of the validity of the ratings of pathologist A, is a single quantity, $\exp(7.74) = 2,298$. This means that among those slides at $\theta_2 = 0$, the odds of rater A classifying a "true +" slide as "+" is 2,298 times as high as classifying a "true –" slide as "+." Similarly, among those

---

[6]For each patient, we can obtain the posterior distribution for the first factor. This posterior distribution can be used determine whether a patient has carcinoma, corrected for rater bias (the second factor).

Table 3.3: Estimates of the Unconditional Latent Class Probabilities and the Conditional Item Probabilities Obtained From the Two-Factor LC Model With the Rater Agreement Data.

		$\theta_1 = 0$ (True $-$)		$\theta_1 = 1$ (True $+$)	
		$\theta_2 = 0$	$\theta_2 = 1$	$\theta_2 = 0$	$\theta_2 = 1$
Class size		0.35	0.18	0.31	0.16
Rater F	$-$	1.00	0.99	0.80	0.11
	$+$	0.00	0.01	0.20	0.89
Rater D	$-$	1.00	0.98	0.61	0.11
	$+$	0.00	0.02	0.39	0.89
Rater C	$-$	1.00	1.00	0.22	0.14
	$+$	0.00	0.00	0.78	0.86
Rater A	$-$	0.94	0.59	0.01	0.00
	$+$	0.06	0.41	0.99	1.00
Rater G	$-$	0.99	0.46	0.01	0.00
	$+$	0.01	0.54	0.99	1.00
Rater E	$-$	0.94	0.28	0.03	0.00
	$+$	0.06	0.72	0.97	1.00
Rater B	$-$	0.87	0.01	0.03	0.00
	$+$	0.13	0.99	0.97	1.00

slides at $\theta_2 = 1$, the corresponding odds ratio is also 2,298.

We could instead express the effect of Factor 1 in terms of differences between probabilities. Such a linear effect is easier to interpret, but is not the same for both types of slides. For slides at $\theta_2 = 0$, the probability of classifying a "true $+$" slide as "$+$" is .94 higher than classifying a "true $-$" slide as "$+$"(.99-.06=.94), whereas for slides at $\theta_2 = 1$, it is .59 higher (1.00 - .41=.59), markedly different quantities. This illustrates that for the linear model, a large interaction term is needed to reproduce the results obtained from the logistic LC model.

Given that a substantial interaction must be added to the linear model to capture the differential biases among the raters, it might be expected that the traditional (linear) FA model also fails to capture this bias. This turns out to be the case, as the traditional rule of choosing the number of factors to be equal to the number of eigenvalues greater than 1 yields only a single factor: The largest eigenvalue was 4.57, followed by 0.89 for the second largest. Despite this result, for purposes of comparison with the LCFA solution, we fitted a two-factor model anyway, using maximum

Table 3.4: Logit and Linearized Parameter Estimates for the Two-Factor LC Model Applied to the Rater Agreement Data.

	Logit		Communality		Linearized		
Rater	θ_1	θ_2	Linear	Total	θ_1	θ_2	$\theta_1\theta_2$
F	7.2	3.4	0.45	0.60	0.53	0.38	0.40
D	6.0	2.6	0.47	0.54	0.62	0.26	0.26
C	7.2	0.5	0.68	0.68	0.82	0.04	0.04
A	7.7	2.4	0.72	0.75	0.82	0.18	-0.16
G	10.1	5.2	0.76	0.82	0.82	0.27	-0.25
E	6.4	3.8	0.65	0.75	0.72	0.35	-0.31
B	5.3	6.3	0.59	0.76	0.60	0.47	-0.42

likelihood for estimation.

Table 3.5 shows that the results obtained from Varimax (orthogonal) and Quartimax (oblique) rotations differ substantially. Hence, without theoretical justification for one rotation over another, FA produces arbitrary results in this example.

The three right-most columns of Table 3.4 present results from a linearization of the LCFA model using the following equation to obtain "linearized loadings" for each variable y_k

$$\widehat{E}(y_k|\theta_1,\theta_2) = b_{k0} + b_{k1}\theta_1 + b_{k2}\theta_2 + b_{k12}\theta_1\theta_2.$$

These three loadings have clear meanings in terms of the magnitude of validity and bias for each rater. They have been used to sort the raters according to the magnitude and direction of bias. The logit loadings do not provide such clearinformation.

The loading on θ_1 corresponds to a measure of validity of the ratings. Raters C, A, and G who have the highest loadings on the first linearized factor show the highest level of agreement among all raters. The loading on θ_2 relates to the magnitude of bias and the loading on $\theta_1\theta_2$ indicates the direction of the bias. For example, in Table 3.3 we saw that raters F and B show the most bias, F in the direction of "false −" ratings and B in the direction of "false +." This is exactly what is picked up by the nonlinear term: the magnitude of the loadings on the nonlinear term (Table 3.4) is highest for these two raters, one occurring as "+," the other as "−."

Table 3.4 also lists the communalities ($R^2_{y_k}$ values) for each rater, and decomposes these into linear and nonlinear portions (the "Total" column includes the sum of the linear and nonlinear portions). The linear portion

Table 3.5: Loadings and Communalities Obtained When Applying a Traditional Two-Factor Model to the Rater Agreement Data.

	Commu-	Varimax		Quartimax	
Rater	nality	θ_1	θ_2	θ_1	θ_2
F	0.49	0.23	0.66	0.55	0.43
D	0.60	0.29	0.72	0.63	0.45
C	0.62	0.55	0.56	0.77	0.18
A	0.73	0.71	0.48	0.85	0.03
G	0.86	0.83	0.42	0.92	-0.09
E	0.78	0.82	0.31	0.86	-0.18
B	0.69	0.80	0.24	0.80	-0.22

is the part accounted for by $b_{k1}\theta_1 + b_{k2}\theta_2$, and the nonlinear part concerns the factor interaction $b_{k12}\theta_1\theta_2$. Note the substantial amount of nonlinear variation that is picked up by the LCFA model. For comparison, the leftmost column of Table 3.5 provides the communalities obtained from the FA model, which are quite different from the ones obtained with the LCFA model.

3.4.2 MBTI Personality Items

In our second example we analyzed 19 dichotomous items from the Myers-Briggs Type Indicator (MBTI) test—seven indicators of the sensing-intuition (S-N) dimension, and 12 indicators of the thinking-feeling (T-F) personality dimension.[7] The total sample size was 8,344. These items were designed to measure two hypothetical personality dimensions, which were posited by Carl Jung to be latent dichotomies. The purpose of the presented analysis was to investigate whether the LCFA model was able to identify these two theoretical dimensions and whether results differed from the ones obtained with a traditional factor analysis.

We fitted zero-, one-, two-, and three-factor models for this data set. Strict adherence to a fit measure like BIC or a similar criterion suggest that more than two latent factors are required to fit these data due to violations of the local independence assumption. This is due to similar wording used in several of the S-N items and similar wording used in some

---

[7]Each questionnaire item involves making a choice between two categories, such as, for example, between thinking and feeling, convincing and touching, or analyze and sympathize.

of the T-F items. For example, in a three-factor solution, all loadings on the third factor are small except those for S-N items S09 and S73. Both items ask the respondent to express a preference between "practical" and a second alternative (for item S09, 'ingenious;' for item S73, "innovative"). In such cases, additional association between these items exists which is not explainable by the general S-N (T-F) factor. For our current purpose, we ignore these local dependencies and present results of the two-factor model.

In contrast to our first example, the decomposition of communalities ($R^2_{y_k}$ values) in the right-most columns of Table 3.6 shows that a linear model can approximate the LCFA model here quite well. Only for a couple of items (T35, T49, and T70) is the total communality not explained to two decimal places by the linear terms only. The left-most columns of Table 3.6 compares the logit and linearized "loadings" ($p_{\theta_\ell y_k}$) for each variable. The fact that the latter numbers are bounded between -1 and $+1$ offers easier interpretation.

The traditional FA model also does better here than the first example. The first four eigenvalues are 4.4, 2.8, 1.1, and 0.9. For comparability to the LC solution, Table 3.7 presents the loadings for the two-factor solution under Varimax (orthogonal) and Quartimax (oblique) rotations. Unlike the first example where the corresponding loadings showed considerable differences, these two sets of loadings are quite similar. The results are also similar to the linearized loadings obtained from the LCFA solution.

The right-most column of Table 3.7 shows that the communalities obtained from FA are quite similar to those obtained from LCFA. In general, these communalities are somewhat higher than those for LCFA, especially for items S27, S44, and S67.

Figure 3.1 displays the two-factor LCFA bi-plot for these data (see Magidson & Vermunt, 2001, 2004). The plot shows how clearly differentiated the S-N items are from the T-F items on both factors. The seven S-N items are displayed along the vertical dimension of the plot which is associated with Factor 2, and the T-F items are displayed along the horizontal dimension, which is associated with Factor 1. This display turns out to be very similar to the traditional FA loadings plot for these data. The advantage of this type of display becomes especially evident when nominal variables are included among the items.

3.4.3   Types of Survey Respondents

We now consider an example that illustrates how these tools are used with nominal variables. It is based on the analysis of four variables from the 1982 General Social Survey (white respondents) given by McCutcheon (1987) to illustrate how standard LC modelling can be used to identify different types

Table 3.6: Logit and Linearized Parameter Estimates and Communalities for the Two-Factor LC Model as Applied to 19 MBTI Items.

	Logit		Linear		Communality	
Item	θ_1	θ_2	θ_1	θ_2	Linear	Total
S02	-0.03	-1.51	-0.01	-0.61	0.37	0.37
S09	-0.01	-1.16	0.00	-0.50	0.25	0.25
S27	0.03	1.46	0.01	0.55	0.30	0.30
S34	-0.07	-1.08	-0.03	-0.45	0.21	0.21
S44	-0.11	1.13	-0.04	0.47	0.22	0.22
S67	-0.06	1.54	-0.02	0.53	0.28	0.28
S73	-0.01	-1.05	0.00	-0.46	0.21	0.21
T06	1.01	0.53	0.43	0.19	0.22	0.22
T29	1.03	0.59	0.44	0.20	0.23	0.23
T31	-1.23	-0.47	-0.52	-0.15	0.29	0.29
T35	-1.42	-0.29	-0.55	-0.09	0.31	0.32
T49	1.05	0.65	0.44	0.22	0.24	0.25
T51	1.32	0.30	0.53	0.09	0.29	0.29
T53	1.40	0.77	0.56	0.22	0.36	0.36
T58	-1.46	-0.12	-0.62	-0.03	0.38	0.38
T66	-1.23	-0.27	-0.54	-0.09	0.30	0.30
T70	1.07	0.61	0.43	0.19	0.22	0.23
T75	-1.01	-0.39	-0.45	-0.14	0.22	0.22
T87	-1.17	-0.45	-0.50	-0.15	0.28	0.28

Table 3.7: Loadings and Communalities From Traditional Factor Analysis of the 19 MBTI Items.

	Quartimax		Varimax		Commu-
	θ_1	θ_2	θ_1	θ_2	nality
S02	0.08	-0.63	0.06	-0.63	0.40
S09	0.07	-0.50	0.06	-0.50	0.26
S27	-0.06	0.62	-0.05	0.62	0.38
S34	0.07	-0.46	0.06	-0.46	0.22
S44	-0.02	0.55	0.00	0.55	0.30
S67	-0.02	0.64	-0.01	0.64	0.41
S73	0.06	-0.46	0.05	-0.46	0.21
T06	-0.49	0.09	-0.49	0.10	0.25
T29	-0.49	0.10	-0.49	0.11	0.25
T31	0.56	-0.04	0.56	-0.05	0.32
T35	0.58	0.05	0.58	0.04	0.34
T49	-0.50	0.13	-0.50	0.15	0.27
T51	-0.57	-0.03	-0.57	-0.02	0.33
T53	-0.61	0.09	-0.61	0.10	0.38
T58	0.64	0.11	0.64	0.10	0.42
T66	0.58	0.05	0.58	0.03	0.33
T70	-0.49	0.10	-0.49	0.11	0.25
T75	0.50	-0.03	0.50	-0.04	0.25
T87	0.55	-0.04	0.55	-0.05	0.30

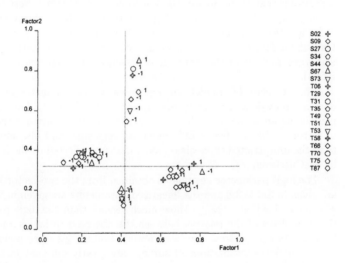

Figure 3.1: Bi-plot of Two-Factor LC Model as Applied to the 19 MBTI Items.

of survey respondents.

Two of the variables ascertain the 1202 respondent's opinion regarding (a) the purpose of surveys (good, depends, or waste of time and money) and (b) how accurate survey are (mostly true or not true), and the others are evaluations made by the interviewer of (c) the respondent's levels of understanding of the survey questions (good, fair/poor) and (d) cooperation shown in answering the questions (interested, cooperative, or impatient/hostile). McCutcheon initially assumed the existence of two latent classes corresponding to 'ideal' and 'less than ideal' types. The purpose of the present analysis is to show how to apply the LCFA model with nominal indicators; that is, to answer the question as to whether these four items measure a single dimension as hypothesized by McCutcheon or whether there are two underlying dimensions. Note that it is not possible to use traditional factor analytic techniques with nominal indicators. A more elaborate analysis of this data set is presented in Magidson and Vermunt (2004).

The two-class LC model—or, equivalently, the one-factor LC model—does not provide a satisfactory description of this data set. Two options for proceeding are to increase the number of classes or to increase the number of factors. The two-factor LC model fits very well, and also much better than the unrestricted three-class model that was selected as the final model by McCutcheon.

The logit parameter estimates obtained from the two-factor LC model are given in Table 3.8 and the linearized parameters are given in Table 3.9. The factor loadings $(p_{\theta_\ell y_k})$ show much clearer than the logit parameters the magnitude of the relation between the observed variables and the two factors. As can be seen, the interviewers' evaluations of respondents and the respondents' evaluations of surveys are clearly different factors. The communalities $(R^2_{y_k})$ reported in the two right-most columns of Table 3.9 show that the linear approximation is accurate for each of the four items.

Figure 3.2 depicts the bi-plot containing the category scores of the four indicators. The plot shows that the first dimension differentiates between the categories of understanding and cooperation and the second between the categories of purpose and accuracy. This display is similar to the plot obtained in multiple correspondence analysis (Van der Heijden, Gilula, & Van der Ark, 1999).

3.5 Conclusion

In this study, we compared LCFA with FA in three example applications where the assumptions of FA were violated. In the MBTI example, the

Table 3.8: Logit Parameter Estimates for the Two-Factor LC Model as Applied to the GSS'82 Respondent-Type Item.

Item	Category	θ_1	θ_2
Purpose	Good	-1.12	2.86
	Depends	0.26	-0.82
	Waste	0.86	3.68
Accuracy	Mostly true	-0.52	-1.32
	Not true	0.52	1.32
Understanding	Good	-1.61	0.58
	Fair/poor	1.61	-0.58
Cooperation	Interested	-2.96	-0.57
	Cooperative	-0.60	-0.12
	Impatient/hostile	3.56	0.69

Table 3.9: Linearized Parameter Estimates and Communalities for the Two-Factor LC Model as Applied to the GSS'82 Respondent-Type Items.

	Loadings		Communalities	
	θ_1	θ_2	Linear	Total
Purpose	0.14	0.45	0.24	0.26
Accuracy	0.15	0.55	0.33	0.33
Understanding	0.57	0.14	0.35	0.36
Cooperation	0.42	0.07	0.18	0.19

resulting linear factor model obtained from standard FA provided results that were quite similar to those obtained with LCFA, although the factors were taken to be dichotomous in the LCFA model. In this case, decomposition of the LCFA solution into linear and nonlinear portions suggested that the systematic portion of the results was primarily linear, and the linearized LCFA solution was quite similar to the FA solution. However, the LCFA model was able to identify pairs and small groups of items that have similar wording because of some violations of the assumption of local independence.

In the rater-agreement example, LCFA results suggested that the model contained a sizeable nonlinear component, and in this case the standard FA was unable to capture differential biases between the raters. Even when a

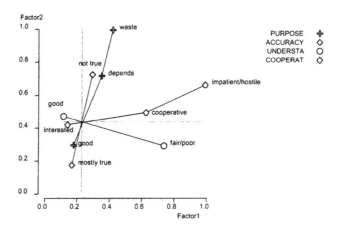

Figure 3.2: Bi-plot of Two-Factor LC Model as Applied to the GSS'82 Respondent-Type Items.

second factor was included in the model, no meaningful interpretation of this second factor was possible, and the loadings from two different rotations yielded very different solutions.

The third example illustrated the used of LCFA with nominal indicators, a situation for which standard FA techniques cannot be used at all. For this example, the factor-analytic loadings and communalities obtained with the proposed linear approximation provided much easier interpretation than the original logit parameters.

Overall, the results suggest improved interpretations from the LCFA approach, especially in cases where the nonlinear terms represent a significant source of variation. This is due to the increased sensitivity of the LCFA approach to all kinds of associations among the variables, not being limited as in the standard linear FA model to the explanation of simple correlations.

The linearized LCFA parameters produced improved interpretation, but in the rater agreement example, a third (nonlinear) component model was needed in order to extract all of the meaning from the results. This current investigation was limited to two dichotomous factors. With three or more dichotomous factors, in addition to each two-way interaction, additional loadings associated with components for each higher-order interaction might also be necessary. Moreover, for factors containing three or more levels, additional terms are required. Further research is needed to explore these issues in practice.

References

Aitkin (1999). A general maximum likelihood analysis of variance components in generalized linear models. *Biometrics, 55*, 218-234.

Agresti, A. (2002). *Categorical data analysis, Second Edition.* New York: Wiley.

Bartholomew, D. J., & Knott, M. (1999). *Latent variable models and factor analysis.* London: Arnold.

Dayton, C. M., & Macready, G. B. (1988). Concomitant-variable latent-class models. *Journal of the American Statistical Association, 83,* 173-178.

Goodman, L. A. (1974). Exploratory latent structure analysis using both identifiable and unidentifiable models. *Biometrika, 61*, 215-231.

Hagenaars, J. A. (1988). Latent structure models with direct effects between indicators: local dependence models. *Sociological Methods and Research, 16,* 379-405.

Hagenaars, J. A. (1990). *Categorical longitudinal data: Log-linear analysis of panel, trend and cohort data.* Newbury Park, CA: Sage.

Heinen, T. (1996). *Latent class and discrete latent trait models: Similarities and differences.* Thousand Oaks, CA: Sage.

Landis, J. R., & Koch, G. G. (1977). The measurement of observer agreement for categorical data, *Biometrics, 33*, 159-174.

Magidson, J., & Vermunt, J. K. (2001). Latent class factor and cluster models, bi-plots and related graphical displays, *Sociological Methodology, 31,* 223-264.

Magidson, J., & Vermunt, J. K. (2004). Latent class models. In D. Kaplan (Ed.), *Handbook of quantitative methods in social science research* (pp. 175-198). Newbury Park, CA: Sage.

McCutcheon, A. L. (1987). *Latent class analysis,* Sage University Paper. Newbury Park, CA: Sage.

Moustaki, I., & Knott, M. (2000). Generalized latent trait models. *Psychometrika, 65,* 391-412.

Reckase, M. D. (1997). A linear logistic multidimensional model for dichotomous item response data. In W. J. van der Linden & R. K. Hambleton (Eds.), *Handbook of modern item response theory* (pp. 271-286). New York: Springer.

Van der Heijden, P. G. M., Dessens, J., & Böckenholt, U. (1996). Estimating the concomitant-variable latent class model with the EM algorithm. *Journal of Educational and Behavioral Statistics, 5,* 215-229.

Van der Heijden P. G. M., Gilula, Z., & Van der Ark, L. A. (1999). On a relationship between joint correspondence analysis and latent class analysis. *Sociological Methodology, 29,* 147-186.

Vermunt, J. K. (2001). The use restricted latent class models for defining and testing nonparametric and parametric IRT models. *Applied Psychological Measurement, 25,* 283-294.

Vermunt, J. K., & Van Dijk, L. (2001). A nonparametric random-coefficients approach: the latent class regression model. *Multilevel Modelling Newsletter, 13,* 6-13.

Vermunt, J. K. (2004). An EM algorithm for the estimation of parametric and nonparametric hierarchical nonlinear models. *Statistica Neerlandica, 58,* 220-233.

Vermunt, J. K., & Magidson, J. (2000). Latent GOLD 2.0 user's guide [Software manual.] Belmont, MA: Statistical Innovations.

Vermunt, J. K., & Magidson, J. (2002). Latent class cluster analysis. In J. A. Hagenaars & A. L. McCutcheon (Eds.), *Applied latent class analysis* (pp. 89-106). Cambridge, UK: Cambridge University Press.

Vermunt, J. K., & Magidson, J. (2003). Addendum to the Latent GOLD user's guide: Upgrade manual for version 3.0 [Software Manual.] Belmont, MA: Statistical Innovations.

Chapter 4

Bayesian Computational Methods for Inequality Constrained Latent Class Analysis

Olav Laudy, Jan Boom, and Herbert Hoijtink
Utrecht University

4.1 Introduction

Exploratory latent class analysis (ELCA) (Clogg, 1981; Goodman, 1974; Haberman, 1988; Vermunt, 1997) is used to group responses x_{ij} of persons $i = 1, \ldots, N$ to items $j = 1, \ldots, J$ into classes $q = 1, \ldots, Q$ such that persons with similar responses are assigned to the same class. In this chapter we restrict ourselves to dichotomous data $x_{ij} \epsilon \{0, 1\}$. Each class q is characterized by J class specific probabilities π_{qj}, indicating the probability of the response '1' on item j in class q, and a weight ω_q, indicating the unconditional probability that a person's latent class membership τ equals q. Let $\mathbf{X} = [\mathbf{x}_1, \ldots, \mathbf{x}_N]$, $\boldsymbol{\theta} = [\boldsymbol{\omega}, \boldsymbol{\pi}_1, \ldots, \boldsymbol{\pi}_Q]$, $\boldsymbol{\pi}_q = [\pi_{q1}, \ldots, \pi_{qJ}]$, $\mathbf{x}_i = [x_{i1}, \ldots, x_{iJ}]$ and $\boldsymbol{\omega} = [\omega_1, \ldots, \omega_Q]$. The density of the data given the parameters of ELCA is then given by

$$P(\mathbf{X}|\boldsymbol{\theta}) = \prod_{i=1}^{N} P(\mathbf{x}_i \mid \boldsymbol{\theta}) = \prod_{i=1}^{N} \left[\sum_{q=1}^{Q} P(\mathbf{x}_i, \tau = q \mid \boldsymbol{\theta}) \right] =$$

$$\prod_{i=1}^{N} \left[\sum_{q=1}^{Q} \omega_q \prod_{j=1}^{J} \pi_{qj}^{x_{ij}} (1 - \pi_{qj})^{(1-x_{ij})} \right]. \tag{4.1}$$

A key question in ELCA is into how many homogeneous subgroups the sample should be divided? Usually fit measures (Everitt, 1988; Lin & Dayton, 1997) are used to determine which number of classes is optimal. Another question concerns the interpretation of the resulting classes. Sometimes classes can be interpreted independent of other classes. As is illustrated in the next section, one class may account for persons with highly developed emotional skills, while an other class accounts for persons with highly developed social skills. It can also be that classes can be ordered with respect to one or more underlying dimensions (Croon, 1990). An example of the latter is an ELCA resulting in three latent classes that can be used to order persons with respect to different levels of social skills (a one-dimensional ordering). It even might be the case that the persons can be ordered with respect to two dimensions, for example, the combinations of levels of social skills and the levels of emotional skills.

A researcher using exploratory analysis behaves as if his research field has not yet been explored very thoroughly, and theories are not yet fully developed. After the execution of an exploratory analysis, a researcher has to determine whether the outcome is in accordance with an existing theory, or that a new theory is emerging. This approach has two drawbacks. First of all, it may not at all be clear which theory corresponds best to the outcomes. This may lead to over-interpretation and guessing. Secondly, scientific progress may be larger if the current state of affairs (existing knowledge and theories) are properly accounted for in the statistical models used for the analyses.

ELCA has been done in areas that have been thoroughly explored, and where theories are well developed, for example, Boom, Hoijtink, and Kunnen (2001) and Jansen and Van der Maas (1997). They use ELCA to analyze data with respect to the Piagetian Balance Scale Task. In section 4.5.1 new data with respect to this task are analyzed using confirmatory latent class analysis (CLCA). There it is also shown how CLCA can be used to refine (the best of) the existing theories, that is, how a new theory can be generated using the old theory as the point of departure.

In this chapter, a specific form of CLCA is proposed (Hoijtink & Molenaar, 1997; Hoijtink, 1998; Hoijtink, 2001). The approach allows a theory to be translated into a CLCA using inequality constraints among the parameters of the model. This can be done for several competing theories. Two fit measures are presented that can be used to select the model that receives the most support from the data.

4.2 Translation of Theories into CLCA

Several models can be constructed using constraints of the following types for $q \neq q'$ and/or $j \neq j'$

$$\pi_{qj} \; > \; \pi_{q'j'},$$
$$\pi_{qj} \; < \; \pi_{q'j'}.$$

To start with a simple example, suppose that persons have to respond to ten items. The first five items can be answered using skills related to social qualities (e.g., do you think you have a good understanding of other people?), the others using skills related to emotional qualities (e.g., do you easily succeed in managing yourself?). The answers to these questions are coded 1 (well-developed) and 0 (undeveloped). Thus, the response vector of each respondent has ten scores with realization 1 or 0. Suppose, several theories exist for these data. A researcher thinks skills related social and emotional are not distinct, leading to the conclusion that there are only two groups of persons: persons who have a higher (social/emotional) intelligence and persons who have a lower intelligence. This *common intelligence theory* can be translated into a latent class model with two latent classes. The class specific probabilities for the first class are all high, the class specific probabilities for the second class are all low, thus meaning that the persons who have both well developed social and emotional skills are allocated in class one, and the less intelligent persons who have both less developed social and emotional skills are allocated in class two. In terms of restrictions (see Table 4.1): For the common intelligence theory, the class specific probabilities of all items in the first class are restricted to be larger than those of all items in the second class. Note that j is used to indicate item numbers, ω_1 denotes the proportion of persons in class 1, and π_{2j} denotes the probability of responding '1' to item j in class 2.

Another researcher might not agree with the common intelligence theory and states that there are indeed two groups of persons, but one group has higher social related skills, while the other group has higher emotional related skills. From this *specific (social/emotional) intelligence theory* it can be inferred that in one class the probabilities of responding 'developed'— that is, the response indicates that the person has well-developed skills—to the social items are higher than for the emotional items, while for the other

Table 4.1: Items and Restrictions on the Response Probabilities for Common (Social/Emotional) Intelligence Theory.

Item type	Items	Restrictions		
Social	1-5	π_{1j}	>	π_{2j}
Emotional	6-10	π_{1j}	>	π_{2j}

Table 4.2: Inequality Constraints for the Specific Intelligence Theory.

Item type	Items	Restrictions		
Social	1-5	π_{1j}	>	π_{2j}
		>		<
Emotional	6-10	π_{1j}	<	π_{2j}

class the probabilities of responding 'developed' to the emotional items are higher than for the social items. This theory can be translated into a CLCA as indicated in Table 4.2: The first five items in the first class have probabilities that are restricted to be larger than the first five items in the second class. The first five items in the first class are also restricted to be larger than the last five items in the first class. The last five items in the second class have probabilities that are restricted to be larger than the last five items in the first class. The last five items in the second class are restricted to be larger than the first five items in the second class.

An alterative display of the inequality constraints for the 'specific intelligence' theory is given in Table 4.3. Here the inequality constraints are implicit; for example, a minus sign indicates a class specific probability is restricted to be smaller than all the class specific probabilities corresponding to a plus sign. This type of display is used in section 4.5.1, where the display with inequality signs is too complicated or impossible. The inequality constraints are implicit: - < +. Note that a minus sign is not restricted with respect to any other minus sign, and a plus sign is not restricted with respect to any other plus sign.

Table 4.3: Alternative Display of the Inequality Constraints for the Specific Intelligence Theory.

Item type	Items	Restrictions	
Social	1-5	+	-
Emotional	6-10	-	+

4.3 Estimates for the CLCA

In this section we explain how estimates of the parameters are obtained. The general algorithm is described by Gelfand, Smith, and Lee (1992) and the direct application to CLCA can be found in Hoijtink (1998). The basic principle is to use the posterior distribution to obtain a sample of the model parameters. This sample can be seen as a discrete representation of the posterior distribution. With this sample, further calculations are easy, for example, the average of the sampled values is the expected a posteriori (EAP) estimate of a parameter, and the 2.5-th and 97.5-th percentile of the sampled values constitute a 95% central credibility interval. Since it is not trivial to obtain a sample from a multivariate posterior distribution, the Gibbs sampler is applied. This algorithm renders a sample from the joint posterior of the parameters by repeatedly sampling from conditional distributions, that is, the distribution of the parameter at hand given all the other parameters.

4.3.1 Posterior Distribution

The density of the data given the parameters of the model is given by Equation 4.1. For each model $k = 1, \ldots, K$, where K denotes the number of models under consideration, the set of inequality constraints is denoted by H_k. The latter will be included in the posterior distribution via the prior distribution. In this chapter, all the priors are chosen to be uniform for all combinations of parameter values allowed by H_k. Note that since information about the models is included in the prior distributions via inequality constraints, in that respect the priors are informative. The conjugate prior for a (constrained) class specific probability is a (truncated) Beta(1,1) distribution. The conjugate prior for the class weights is a Dirichlet distribution parameterized such that a priori all combinations of weight values summing to one are equally likely, that is, Dirichlet$(\alpha_1, \ldots, \alpha_Q)$, with $\alpha_q = 1$. The resulting posterior $P(\theta \mid \mathbf{X}, H_k)$ is proportional to the product of the density of the data $P(\mathbf{X} \mid \theta)$ and the (truncated) proportional prior

$P(\boldsymbol{\theta} \mid H_k)$, that is

$$P(\boldsymbol{\theta} \mid \mathbf{X}, H_k) \propto P(\mathbf{X} \mid \boldsymbol{\theta}) \times P(\boldsymbol{\theta} \mid H_k),$$

where $P(\boldsymbol{\theta} \mid H_k)$ has the value 1 if $\boldsymbol{\theta}$ is in accordance with the constraints imposed by H_k, and 0 otherwise.

4.3.2   Gibbs Sampler

The Gibbs sampler is an iterative procedure. In iteration $r = 0$ initial values have to be provided for the class weights and the class specific probabilities. Any set of values that is in agreement with the constraints imposed upon the parameters can be used. Each iteration $r = 1, \ldots, R$ consists of three steps:

Step 1: For $i = 1, \ldots, N$, sample class membership $\tau_{i,r} \in \{1, \ldots, Q\}$ from its posterior distribution given the current values (i.e., the values sampled in iteration $r-1$) of the class weights, the class specific probabilities and the data. This conditional posterior is a Multinomial distribution with probabilities

$$P(\tau_{i,r} = q \mid \mathbf{x}_i, \boldsymbol{\theta}_{r-1}) = \frac{P(\mathbf{x}_i, \tau_{i,r} = q \mid \boldsymbol{\theta}_{r-1})}{P(\mathbf{x}_i \mid \boldsymbol{\theta}_{r-1})} \tag{4.2}$$

for $q = 1, \ldots, Q$. Note that both the numerator and the denominator in the right-hand side of Equation 4.2 are defined in Equation 4.1.

Step 2: For $q = 1, \ldots, Q$ and $j = 1, \ldots, J$, sample π_{qj} from its posterior distribution given the current values of τ_i for $i = 1, \ldots, N$, and the data and the constraints. This conditional posterior is a (truncated) Beta distribution with parameters $s_{qj,r} + 1$ and $N_{q,r} - s_{qj,r} + 1$, where $N_{q,r}$ denotes the number of persons allocated to class q in iteration r, and $s_{qj,r}$ denotes the number of persons allocated to class q in iteration r that respond 1 to item j. Note that the Beta distribution is truncated because the sampled value for π_{qj} is only acceptable if it is in accordance with the inequality constraints involving π_{qj}. The naive way to do so is: Sample from the correct (non-truncated) Beta distribution until a deviate is sampled that satisfies the constraints. However, this is quite inefficient when only a small range of the distribution is admissible. Inverse probability sampling solves this problem. Let π_{qj} be the parameter that has to be sampled from the truncated Beta distribution. The lower bound a is given by the largest class specific probability that, according to the constraints imposed by the model at hand, must be smaller than π_{qj}. The upper bound b is the smallest class specific probability that, according to the constraints imposed by the model at hand, must be greater than π_{qj}. The sampling is achieved using a uniform (0,1) deviate U and the computation of

$$\pi_{qj} = \Phi_{\pi_{qj}}^{-1} \{ \Phi_{\pi_{qj}}(a) + U[\Phi_{\pi_{qj}}(b) - \Phi_{\pi_{qj}}(a)] \},$$

where $\Phi_{\pi_{qj}}(a)$ is the proportion of the conditional posterior distribution (a truncated Beta distribution) of π_{qj} below a and $\Phi_{\pi_{qj}}(b)$ is the proportion of conditional posterior distribution below b. $\Phi_{\pi_{qj}}^{-1}\{.\}$ denotes the inverse cumulative density evaluated at the argument. This procedure always renders a deviate from the conditional distribution at hand within the bounds a and b (Gelfand et al., 1992).

Step 3: Sample the class weights from their posterior distribution given the current values of τ_i for $i = 1, \ldots, N$. This posterior is a Dirichlet distribution with parameters $N_{1,r} + 1, \ldots, N_{Q,r} + 1$.

For all analyses executed in the chapter, the Gibbs sampler was run for 110,000 iterations. After a burn-in period of 10,000 iterations the values sampled in the second and third step of each 100-th iteration were saved (these iterations are denoted using the superscript $m = 1, \ldots, M$). The result is $\theta^1, \ldots, \theta^m, \ldots, \theta^{1,000}$. This sample can be used to obtain estimates of the model parameters and the corresponding credibility intervals, taking into account the prior constraints. The expected a posteriori (EAP) estimate of a parameter is simply the average of the 1,000 values of that parameter sampled from the posterior distribution. A 95% central credibility for this parameter is given by the 2.5-th and 97.5-th percentile of the distribution of these 1,000 sampled values. In the next section it is shown that it is easy to compute and evaluate fit measures using the sample from the posterior distribution.

4.4   Model Selection

After the translation of a number of competing theories into constrained latent class models, the support the data provide for each latent class model has to be determined. Three fit measures that can be evaluated using Bayesian computational methods (the marginal likelihood, posterior model probabilities, and the pseudo likelihood ratio test) have been proposed in the literature (Kass & Raftery, 1995; Hoijtink, 2001). For a discussion of the performance of these measures in the context of inequality constrained models, the interested reader is referred to Hoijtink (1998, 2001). These fit measures are discussed in the next sections.

4.4.1 Marginal Likelihood and Posterior Model Probabilities

Kass and Raftery (1995) present a comprehensive review of the marginal likelihood and posterior probability of a model. The basic idea behind the marginal likelihood factors is the same as the basic idea behind more familiar information criteria like AIC, CAIC, and BIC. It can, for example, be shown (see Kass & Raftery, 1995), that the Bayesian Information Criterion (Schwarz, 1978) is an approximation of minus twice the logarithm of the marginal likelihood. Although not explicit in its formulation, the marginal likelihood, like the information criteria, contains a trade off between the likelihood of the parameters given the data and the number of parameters in the model.

In the remainder of this chapter, minus twice the logarithm of the marginal likelihood is used

$$-2\log P(\mathbf{X}|H_k) = -2\log \int_{\boldsymbol{\theta}_k} P(\mathbf{X} \mid \boldsymbol{\theta}_k)P(\boldsymbol{\theta}_k \mid H_k)\, d\boldsymbol{\theta}_k, \qquad (4.3)$$

which brings comparisons of different models on the same scale as the familiar deviance statistics (Kass & Raftery, 1995). Loosely formulated, minus twice the logarithm of the marginal likelihood can be interpreted as the distance between the model at hand and the true model: The smaller its value, the smaller the distance.

There are many ways to compute Equation 4.3. In this chapter the method proposed by Kass and Raftery (1995) is used. They suggest to sample 99% of the parameter vectors (in our case 990) from the posterior distribution parameter vectors, and to imagine that 1% of the parameter vectors (in our case 10) is sampled from an imaginary distribution where for each θ $P(\mathbf{X} \mid \theta)$ is equal to the marginal likelihood. An approximation of $-2\log P(\mathbf{X} \mid H_k)$ is denoted as $-2\log \hat{P}$ and obtained via a simple iterative algorithm based on the implicit equation

$$-2\log \hat{P} = -2\log \left[\frac{10\hat{P} + \sum_{m=1}^{990} \frac{P(\mathbf{X}|\theta_k^m)}{.01+P(\mathbf{X}|\theta_k^m)/\hat{P}}}{\hat{P} + \sum_{m=1}^{990} \frac{1}{.01+P(\mathbf{X}|\theta_k^m)/\hat{P}}} \right].$$

If the prior probabilities of the K models under investigation are equal, that is, $P(H_k) = 1/K$ for $k = 1,\ldots,K$, the posterior probability of each model can be computed as

$$P(H_k|\mathbf{X}) = \frac{P(\mathbf{X}|H_k)}{\sum_{k=1}^{K} P(\mathbf{X}|H_k)},$$

for $k = 1, \ldots, K$. The posterior model probability $P(H_k|\mathbf{X})$ denotes the support for model k in the total set of K models given by the data. In this chapter, both the marginal likelihood and the posterior probability of a model are reported.

4.4.2   Pseudo Likelihood Ratio Test

Hoijtink (2001) shows that the likelihood ratio test (Everitt, 1988; Lin & Dayton, 1997) is not performing very well if the goal is to select the best of a number of inequality constrained models. The performance is much better if the pseudo likelihood ratio statistic is used. This statistic is denoted by $D_k(\mathbf{X}, \boldsymbol{\theta}_k)$ and compares for each pair of items, the expected number of each possible pair of responses (i.e., 00, 10, 01, and 11, respectively) to the corresponding observed number. Let $\mathbf{n}_{gh}^{vw}$ denote for items g and h the observed frequencies of the response pattern $X_g = v, X_h = w$ where $v, w \in \{0, 1\}$. Furthermore, let $\mathbf{m}_{gh|k}^{vw}$ denote the expected frequencies of these response patterns given $\boldsymbol{\theta}_k$. The pseudo likelihood ratio statistic is then defined as

$$\mathrm{D}_k(\mathbf{X}, \boldsymbol{\theta}_k) = -2 \sum_{g=1}^{J-1} \sum_{h=g+1}^{J} \sum_{v=0}^{1} \sum_{w=0}^{1} \mathbf{n}_{gh}^{vw} \log \frac{\mathbf{m}_{gh|k}^{vw}}{\mathbf{n}_{gh}^{vw}}. \qquad (4.4)$$

The expected frequencies conditional on $\boldsymbol{\theta}_k$ are computed using

$$\mathbf{m}_{gh|k}^{vw} = N \sum_{q=1}^{Q} \omega_q \left[\pi_{qg}^{v} (1 - \pi_{qg})^{1-v} \right] \left[\pi_{qh}^{w} (1 - \pi_{qh})^{1-w} \right].$$

The larger $D(\mathbf{X}, \boldsymbol{\theta}_k)$, the larger the discrepancy between the data $\mathbf{X}$ and model k.

Because $\boldsymbol{\theta}_k$ is unknown, Equation 4.4 cannot be computed. The classical solution is to substitute the unknown quantity with the maximum likelihood estimate of $\boldsymbol{\theta}_k$. The Bayesian solution uses the posterior distribution of $\boldsymbol{\theta}_k$ (Rubin, 1984; Meng, 1994; Gelman, Meng, & Stern, 1996). The posterior distribution summarizes the available information with respect to $\boldsymbol{\theta}_k$. The posterior can accurately be represented using a sample $\boldsymbol{\theta}_k^1, \ldots, \boldsymbol{\theta}_k^m, \ldots, \boldsymbol{\theta}_k^{1000}$ from this posterior. Each of the 1000 vectors $\boldsymbol{\theta}_k^m$ can be used to generate a replicated data matrix $\mathbf{X}_k^m$ that is in accordance with model k. The procedure is simple: For N persons class membership is sampled from a multinomial distribution with probabilities $\boldsymbol{\omega}_k^m$. Subsequently, the class specific probabilities $\pi_{q1,k}^m, \ldots, \pi_{qJ,k}^m$ of the class to which a person is assigned are compared with a vector of pseudo random numbers sampled from a U(0,1) distribution. If a class specific probability is larger than the corresponding random number, a person gives the response 1, otherwise the response 0 is given. This procedure is repeated for $m = 1, \ldots, 1000$.

For each $\boldsymbol{\theta}_k$ two discrepancies can be computed: $D(\mathbf{X}, \boldsymbol{\theta}_k^m)$, which is a discrepancy between the observed data and the model; and, $D(\mathbf{X}_k^m, \boldsymbol{\theta}_k^m)$, which is a discrepancy between replicated data and the model. If $D(\mathbf{X}_k^m, \boldsymbol{\theta}_k^m)$ $\geq D(\mathbf{X}, \boldsymbol{\theta}_k^m)$, the discrepancy between the observed data and the model is equal to or smaller than the discrepancy between the replicated data and the model. The posterior predictive p-value is the proportion of 1000 comparisons for which $D(\mathbf{X}_k^m, \boldsymbol{\theta}^m) \geq D(\mathbf{X}, \boldsymbol{\theta}^m)$. The posterior predictive p-value is formally defined as

$$p_k = Pr[D(\mathbf{X}_k, \boldsymbol{\theta}_k) \geq D(\mathbf{X}, \boldsymbol{\theta}_k) \mid \mathbf{X}, H_k],$$

that is, the probability that the discrepancy between model k and a data matrix $\mathbf{X}_k$ generated using model k is equal to or larger than the discrepancy between model k and the observed data matrix $\mathbf{X}$. The p_k is an absolute fit measure, that can be used to test the pseudo likelihood ratio statistic, that is, which can be used to determine whether model k accurately describes the data. Stated otherwise, analogous to the interpretation of classical p-values, values smaller than .05 indicate a lack of fit of the model, and values larger than .05 indicate that the model at hand was able to accurately reproduce the observed frequencies.

4.5   Strategies to Solve the Piagetian Balance Scale Task

The balance scale task was recognized in the early eighties as a way of eliciting different rule-governed response patterns for proportionality reasoning (Siegler, 1981). A picture of a simplified balance scale is shown to children. While the beam is fixed, a number of identical weights are placed on each side at certain distances from the fulcrum. For each of a number of balances (the items) the children have to predict which side will tip, if any. The weights on the balance differ with respect to their number and distance to the center. The formal rule to obtain the correct answer is that the balance is in equilibrium when the product of the number of weights and the distance from the center is equal at both sides of the balance.

Applications of ELCA in the context of the balance task can be found in Boom et al. (2001) and Jansen and Van der Maas (1997). New balance scale data will be used to determine which of the existing theories that explain children's responses to the items of the balance scale task is the best. As the result is not conclusive, the best of these theories will be used as the point of departure for further theoretical developments.

Nearly 900 randomly selected Dutch children from 4- to 16-years-old participated, with a mean age of 10.35 (standard deviation 2.82). The children were tested individually at home by students and did not receive feedback until the task had been finished. The assessment was part of a training procedure for psychology students. The students were prepared for this specific assessment in small groups and had to follow a strict assessment protocol.

4.5.1 Theories and Hypotheses About the Data

Siegler (1981) distinguished six types of problems. In *balance* problems, weight and distance are equal on both sides. In *weight* problems, the distance is the same on both sides but the number of weights is different. These first two problem types were not used in this study, since they do not differentiate between the postulated rules and were expected to be answered correctly by all children. In *distance* problems, the weight is the same on both sides but the distance is different. In conflict problems, more weight is on one side and greater distance on the other, such that the side with more weight falls (*conflict-weight* problem), the side with the greater distance falls (*conflict-distance* problem), or the balance remains horizontal (*conflict-balance* problem).

Siegler (1981) described four strategies or rules. Each of these strategies can be characterized by a specific pattern of scores on the different item types.

rule 1 Children will only consider the number of weights on each arm. Therefore it can be expected that they have a higher probability of correctly responding to the weight and the conflict-weight items than to the other item types.

rule 2 Children get a grasp of distance: When the number of weights is equal on both sides, they judge the influence of distance correctly, otherwise they ignore distance and only consider the number of weights. For this strategy, it can be predicted that children have a higher probability of correctly responding to the weight, distance, and conflict-weight items than the other item types.

rule 3 Children will evaluate both the distance and the number of weights correctly, but if one side has more weights and the other side more distance they will be confused and guess. The probability of success will be at chance level (they make a random prediction) for all conflict type of problems.

Table 4.4: Inequality Constraints for Siegler's Model (for Explanation of the Notation, See Text).

Item type	Items	Restrictions			
		Rule 1	Rule 2	Rule 3	Torque
Distance	1-5	-	+	+	+
Conflict Weight	6-9	+	+	±	+
Conflict Distance	10-14	-	-	±	+
Conflict Balance	15-19	-	-	±	+

rule 4 Children will apply the correct (torque) rule. The probability of responding correctly is high for all item types.

As can be seen in Table 4.4, the test used in this chapter contains 19 items of the following types: five distance, four conflict-weight (originally 5 but one item had a printing error in the test booklet for half of the sample), five conflict-distance, and five conflict-balance. In Table 4.4 a translation of Siegler's model into CLCA is elaborated upon. Note that the inequality constraints are implicitly shown: - indicates a low probability of correctly responding to the item, + a high probability of correctly responding to an item, and ± indicates a random prediction. All the probabilities associated with the - signs have to be smaller than the probabilities associated with the ± signs, and all the probabilities associated with the + signs have to be larger than probabilities associated with ±.

Wilkenings and Anderson (1982) argue the existence of another strategy. The *addition rule* suggests that the number of weights and the number of distance intervals on the left are summed and compared to the sum of weights and distances on the right: the side with the greater sum is expected to tip. For the existing item types, we designed the items such that the addition rule could be detected because two conflict-weight items and two conflict-balance items evoke an incorrect response whereas the remaining conflict-weight and conflict-balance and all conflict-distance items evoke a correct response when this rule is applied to this set of items. Children applying the addition rule will have a low probability of correctly responding to the items that evoke an incorrect response, and a high probability of correctly responding to the remaining conflict-items. Normandeau, Larivee, Roulin, and Longeot (1989) argue that rule 3 of Siegler is not homogeneous. Their paper supports the existence of the addition rule and they introduce yet another rule: *qualitative proportion rule*. Children using this rule understand that more weights at a small distance from the fulcrum compensates

Table 4.5: Inequality Constraints for Normandeau's Model (for Explanation of the Notation, See Text).

Item Type	Items	Restrictions				
		Rule 1	Rule 2	Add	QP	T
Distance	1-5	-	+	+	+	+
Conflict Weight	6,9	+	+	-	-	+
Conflict Weight Add	7,8	+	+	+	-	+
Conflict Distance	–	-	-	-	-	+
Conflict Distance Add	10-14	-	-	+	-	+
Conflict Balance	16,19	-	-	-	+	+
Conflict Balance Add	15,17,18	-	-	+	+	+

Note. T = Torque.

for fewer weights at a far distance, resulting in a prediction of balance for all conflict problems. Thus, the qualitative proportion rule predicts that all conflict-weight and conflict-distance items have low probabilities of being answered correctly, and all the conflict-balance items have a high (or higher) probability of a correct response. The five resulting latent classes are displayed in Table 4.5. Note that this table is comparable to Table 4.4, but extended to differentiate between the addition and non-addition items. Moreover, there can be seen that rule 3 has been split up into a latent class accounting for the addition rule and a latent class accounting for the qualitative proportion rule. In the current item set all conflict distance items were solvable using the addition rule.

4.5.2   Results

The model selection measures have been computed for Siegler's model and for Normandeau's model. Note that the pseudo likelihood p-value indicates the absolute fit of the model. A p-value smaller than 0.05 indicates a lack of fit of the model, whereas the p-value is larger than 0.05, the model accurately reproduces the observed frequencies. The marginal likelihood can be interpreted as the distance between the model at hand and the true model: the smaller the value, the smaller the distance. The value of the marginal likelihood is on the same scale as the familiar deviance statistics. Two or more models can be compared using the value of the marginal likelihood. Since the marginal likelihood implicitly uses a parameter penalty, the model with the smallest marginal likelihood value has to be preferred. The marginal likelihoods of all models analyzed can also be transformed

Table 4.6: Fit Evaluation of the Siegler's Model and Normandeau's Model for the Balance Scale Data.

Model	Pseudo Likelihood	-2log Marginal Likelihood	Posterior Model Probability
Siegler	0.003	13421	0
Normandeau	0.019	13206	1.0

into the posterior model probability. This number indicates the support for each model in the total set of models given the data.

In Table 4.6, it can be seen from both the marginal likelihood and the posterior model probability that Normandeau's model is superior. However, as indicated by the p-value of the pseudo likelihood ratio test (smaller than .05), it is questionable whether Normandeau's model adequately reproduces the frequencies with which the response vectors are observed. This lack of fit could be due to existing strategies that are not predicted by the theory.

Figure 4.1 presents the class specific probabilities of Normandeau's model. On the horizontal axis the items are displayed, on the vertical axis the class specific probabilities. Classes one and two clearly represent rule 1 (only considering weight leads to a high probability of answering conflict-weight items (6-9) correct) and rule 2 (high probability of answering conflict-weight (6-9) and distance items (1-5) correct). These rule are dominant, since a substantial part of the sample belongs to these classes. The third class represents the addition rule, although the probabilities for items 15, 17 and 18 are lower than expected for this rule.

Class four represents the qualitative proportion rule. As can be seen only a small proportion of the children belong to this class. Furthermore, although the class specific probabilities are in accordance with the constraints, especially the probabilities for the first and the last five items should have been higher to obtain a convincing representation of a qualitative proportion rule. It can be a 'true' strategy, but maybe it should be specified in more detail than simply by "all conflict items except conflict balance items are answered incorrectly." It could be, for example, that children do have an intuitive idea how distance and weight work together, but only when there is a large difference between the products of weight and distance on both sides.

Knowing that there are very few children that can actually solve the balance scale task, a class size of 29% for rule 4 seems extremely large.

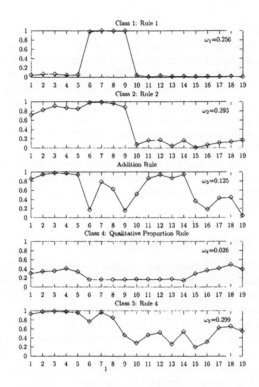

Figure 4.1: Class-Specific Probabilities of the Normandeau Theory.

Furthermore, the class specific probabilities for this correct strategy are all predicted to be high, but as can be seen in the figure, this is not the case. Stated otherwise, class five does not yet give a convincing representation of rule 4.

The results for Normandeau's model are not conclusive. The p-value of the pseudo likelihood ratio test indicates that the data are not adequately reproduced. Furthermore, for classes four and five the class specific probabilities do not give a clear representation of the presumed underlying rule.

It could have been an option to represent the torque rule by a class for which $\pi_{qj} > .90$ for $j = 1, \ldots, J$. However, this value of 0.90 seems rather arbitrary. From the theory it can be predicted that the probability of a correct response in class five has to be higher than the probabilities associated with a random prediction. Note that the method of testing models via the incorporation of inequality constraints on the model parameters explicitly shows that one chooses the best theory from a set of reasonable theories. This means that not all possible models are included, nor guarantees this procedure that the 'true' model is in the set. In the next section, it is shown how this best theory can be used as the point of departure for theory refinement.

4.5.3   Theory Refinement

In this section, Normandeau's model is extended with one and two unconstrained classes, respectively. This constitutes an example of scientific exploration using the current state of knowledge as the point of departure. Note that the results of this exploration are indefinite. To confirm the exploratory results, these findings have to be translated into inequality constraints and they have to be analyzed using new data.

As can be seen in Table 4.7, the Normandeau model with two unconstrained classes receives the most support from the data. Note, that the p-value of the pseudo likelihood ratio test now indicates that the data are adequately reproduced by this model. Furthermore, comparing the posterior probabilities of the three Normandeau models, it is clear that the variant with two extra unconstrained classes is superior.

As can be seen in Figure 4.2, the interpretation of the first four classes is similar to the interpretation given for the Normandeau model without extra classes. Note, however, that the probabilities for items 17 and 18 in class three have increased, that is, class three gives a better representation of the addition rule. The same holds for the first and last five items in class four, which now give a better representation of the Qualitative Proportion rule. Class five now represents rule 4 and contains only a small proportion of the children (as is expected).

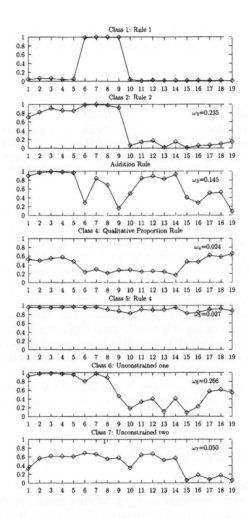

Figure 4.2: Class-Specific Probabilities of the Normandeau Theory Extended With Two Classes.

Table 4.7: **Refined Fit Evaluation of Siegler's Model, Normandeau's Model, and Two Extended Models for the Balance Scale
Data.**

Model	Pseudo Likelihood	-2log Marginal Likelihood	Post
Siegler	0.003	13421	-
Normandeau	0.019	13206	0.0
Normandeau + 1 class	0.054	12909	0.0
Normandeau + 2 classes	0.082	12837	1.0

Class 6 accounts for a fairly large proportion of persons. This class is
difficult to associate with a strategy or rule. Our best guess is that it is
class of children who are somewhere between rule 2 and the addition rule.
The second unrestricted class (class 7) is a global pattern, only grouping
children that have in common that they do not consider the answer 'balance'
an option (the last five items are almost never correctly answered). This
class was also mentioned by Jansen and van der Maas (1997).

In the current version of the balance scale task, the items are chosen
on the basis of being of a certain type (e.g., conflict balance item). The
magnitude of the physical quantities is not varied systematically. We suggest that in further research one chooses the items of the same type more
carefully, such that the role of the physical quantities can be asserted. For
example, choose addition items within the conflict distance item such that
the items vary from a big difference between addition torque to a small
difference in a controlled way.

4.6   Conclusion

This chapter illustrated that theories can be included in latent class models
using inequality constraints among the class specific probabilities. An example from the domain of developmental psychology was used to illustrate
the resulting CLCA. If, in a certain research domain, one or more theories
exist, CLCA has advantages over ELCA. First of all it provides a straightforward way to select the best of a number of competing theories. Secondly,
as illustrated using the balance scale data, it allows theory refinement using
the current state of knowledge as the point of departure.

Siegler and Chen (2002) mention some disadvantages of the LCA. One
is the arbitrary alpha level of 5% and the unclear interpretation of it. This

is acknowledged in Bayesian statistics for quite some time, and a solution has been found in the form of posterior model probability (Sellke, Bayarri, & Berger, 2001). We use this solution, because instead of a probability of incorrectly rejecting the null hypothesis, the posterior model probability gives the probability of the data given the model among other models.

Currently user friendly software containing an implementation of the proposed approach is being developed. Readers interested in this software can send an e-mail to the first author at *o.laudy@fss.uu.nl*. The e-mail should include a description of the research at hand, the data, and the theories involved.

References

Boom, J., Hoijtink, H., & Kunnen, S. (2001). Rules in the balance: Classes, strategies or rules for the balance scale task. *Cognitive Development, 16*, 717-735.

Clogg, C. C. (1981). New developments in latent structure analysis. In D. J. Jackson & E. F. Borgatta (Eds.), *Factor analysis and measurement in sociological research* (pp. 215-246). Beverly Hills, CA: Sage.

Croon, M. A. (1990). Latent class analysis with ordered latent classes. *British Journal of Mathematical and Statistical Psychology, 43*, 171-192.

Everitt, B. S. (1988). A Monte Carlo investigation of the likelihood ratio test for number of classes in latent class analysis. *Multivariate Behavioral Research, 23*, 531-538.

Gelfand, A. E., Smith, A. F. M., & Lee, T. (1992). Bayesian analysis of constrained parameter and truncated data problems using Gibbs sampling. *Journal of the American Statistical Association, 87*, 523-532.

Gelman, A., Meng, X. L., & Stern, H. S. (1996). Posterior predictive assessment of model fitness via realized discrepancies (with discussion). *Statistica Sinica, 6*, 733-807.

Goodman, L. A. (1974). The analysis of systems of qualitative variables when some of the variables are unobservable. Part I: A modified latent structure approach. *American Journal of Sociology, 79*, 1179-1295.

Haberman, S. J. (1988). *Analysis of quantitative data, vol 2. New developments*. New York: Academic Press.

Hoijtink, H., & Molenaar, I. W. (1997). A multidimensional item response model: Constrained latent class analysis using the Gibbs sampler and posterior predictive checks. *Psychometrika, 62*, 171-180.

Hoijtink, H. (1998). Constrained latent class analysis using the Gibbs sampler and posterior predictive p-values: Applications to educational testing. *Statistica Sinica, 8*, 691-711.

Hoijtink, H. (2001). Confirmatory latent class analysis: Model selection using Bayes factors and (pseudo) likelihood ratio statistics. *Multivariate Behavioral Research, 16*, 717-735.

Jansen, B. R. J., & Van der Maas, H. L. J. (1997). Statistical test of rule assessment methodology by latent class analysis. *Developmental Review, 17*, 321-357.

Lin, T. H., & Dayton, C. M. (1997). Model selection information criteria for non nested latent class models. *Journal of Educational and Behavioral Statistics, 22*, 249-264.

Kass, R. E., & Raftery, A. E. (1995). Bayes factors. *Journal of the American Statistical Association, 90*, 773-795.

Meng, X. L. (1994). Posterior predictive p-values. *The Annals of Statistics, 22*, 1142-1160.

Normandeau, S., Larivee, S., Roulin, J., & Longeot, F. (1989). The balance-scale dilemma: Either the subject or the experimenter muddles through. *Journal of Genetic Psychology 150*, 237-250.

Rubin, D. B. (1984). Bayesianly justifiable and relevant frequency calculations for the applied statistician. *The Annals of Statistics, 12*, 1151-1172.

Schwarz, G. (1978). Estimating the dimension of a model. *The Annals of Statistics, 6*, 461-464.

Sellke, T., Bayarri, M. J., & Berger, J. O. (2001). Calibration of p-values for testing precise null hypotheses. *American Statistician 55*, 62-71.

Siegler, R. S. (1981). Developmental sequences within and between concepts. *Monographs of the Society for Research in Child Development. 46*(2, Serial No. 189).

Siegler, R. S., & Chen, Z. (2002). Development of rules and strategies: Balancing the old and the new. *Journal of Experimental Child Psychology, 81*, 446-457.

Vermunt, J. K. (1997). *Log-linear models for event histories.* Thousand Oaks, CA: Sage.

Wilkenings, F., & Anderson, N. H. (1982). Comparison of two rule-assessment methodologies for studying cognitive development and knowledge structure. *Psychological Bulletin, 92*, 215-237.

Chapter 5

Analyzing Categorical Data by Marginal Models

Wicher P. Bergsma
EURANDOM, Eindhoven University of Technology

Marcel A. Croon
Tilburg University

5.1 Introduction

The log-linear model plays a prominent role in the statistical analysis of categorical data. Log-linear analysis of a contingency table aims at obtaining a parsimonious model for the cell probabilities (or expected frequencies) that provides a statistically acceptable explanation of the association among the observed variables with as few parameters as necessary. In general log-linear analysis is applied to the contingency table that contains the observed frequencies from the joint distribution of all variables involved in the analysis. However, many other substantively interesting questions may be asked that pertain not to the original table, but to subsets of marginal tables that can be derived from it. Some of these questions may be answered by a log-linear analysis on these marginal tables. In longitudinal research, for example, changes in the marginal distribution of some variables may be studied by testing hypotheses of marginal homogeneity. Furthermore, changes in the association among variables over time may be studied by testing hypotheses about equality of corresponding interaction parameters in log-linear models for different marginal tables. Some other questions, however, cannot be reformulated in terms of log-linear models. As an example of this type of research question we consider the question of whether the values of some well-known association coefficients change over time.

In this chapter we discuss how all analyses referred to above can be carried out within a unified framework that generalizes the basic log-linear modelling approach. In this framework each model is specified by means of the constraints it imposes on the cell probabilities in a multidimensional contingency table. The maximum likelihood estimates of the constrained cell probabilities and the associated tests of significance are obtained by the procedures developed by Aitchinson (1962), and Aitchinson and Silvey (1958, 1960). In representing these constrained models a notational system is used that is a generalization of that developed by Lang (1996) and Bergsma (1997), which was based on earlier work by Grizzle, Starmer, and Koch (1969), Forthofer and Koch (1973), and Kritzer (1977). In a previous article (Croon, Bergsma, & Hagenaars, 2000) we discussed how this approach could be helpful to test several hypotheses about change in discrete variables over time. A similar set of research problems was considered by Vermunt, Rodrigo, and Ato-Garcia (2000) but their approach is different from ours by requiring the direct estimation of log-linear parameters. Moreover, the scope of these two papers was limited to those hypotheses that could be formulated in terms of the parameters of log-linear models for marginal tables. In this chapter we show how hypotheses that cannot be stated as log-linear models can be tested by specifying the models in terms of the nonlinear inequality constraints they impose on the cell probabilities.

We illustrate our discussion with analyses on the data given in Table 5.1. These data were obtained in a longitudinal survey in which a panel of respondents was interviewed in September and October of the same year on their opinion on the Supreme Court candidate Clarence Thomas (CT) who was nominated by President George Bush Senior. This nomination stimulated some public debate because of CT's Afro-American background and his allegedly extremely conservative views. Moreover, after the first survey was held on September 3-5, a charge for sexual harassment was brought against CT by his former aide Anita Hill on September 25. The second survey was then conducted on October 9. Table 5.1 cross-classifies three variables constructed from these surveys. The variables S and O refer to the respondents' opinions on CT in September and October, respectively. For the present analyses their responses were coded as: 1 = Favorable, 2 = Not Favorable, 3 = Undecided or Haven't heard enough. We refer to this third response as the No clear opinion response alternative. The variable P refers to the respondents' political orientation. We have grouped the original responses to this variable in three categories as follows: 1 = Liberal,

Table 5.1: Cross Classification of Opinions on Clarence Thomas in September (S) and October (O) (1 = Favorable, 2 = Not Favorable, 3 = No Clear Opinion) Against Political Orientation (P with 1 = Liberal, 2 = Moderate, 3 = Conservative).

			P	
S	O	1	2	3
1	1	18	77	99
	2	4	7	3
	3	7	31	21
2	1	4	12	8
	2	43	36	11
	3	7	12	4
3	1	43	108	55
	2	32	41	24
	3	54	127	59

2 = Moderate, 3 = Conservative.

The theoretical probabilities corresponding to the cells of Table 5.1 will be denoted by π_{ijk}^{SOP}. In what follows we often assume that these cell probabilities are assembled in a vector π^{SOP} with each element in this vector corresponding to a particular response pattern on the variables S, O and P. In stacking the elements of a multidimensional table in a vector, we will always work under the convention that the subscripts of the later variables change the fastest. Note also that the symbols for the cell probabilities all have the superscript SOP. When we would only discuss analyses on 5.1, these superscripts are redundant and could be omitted. However, later, when discussing simultaneous analyses on various marginal tables derived from the same table, we need a flexible notation to make clear to which marginal table a particular probability or observed frequency refers. A notation that makes use of superscripts fulfills this need. We make use of a similar notational principle to describe the log-linear parameters in the log-linear models for these tables.

In section 5.2 we introduce our approach by showing how log-linear models for contingency tables can be tested by a constrained estimation of the theoretical cell probabilities. This approach is based on the observation that each log-linear model can be characterized by the constraints it imposes on these cell probabilities. In section 5.3 we extend this approach to the simultaneous log-linear analysis of several marginal tables defined on

the original contingency table. In section 5.4 we describe a further generalization of it for analyzing marginal tables according to models that cannot
be cast into a log-linear form. Section 5.5 discusses the more technical
aspects of the optimization algorithm used in fitting the models.

5.2   Log-Linear Models for a Total Table

For Table 5.1, the most general decomposition of the log probabilities is
given by

$$
\begin{aligned}
\log \pi_{i\,j\,k}^{SOP} \;=\; & \lambda_{*\,*\,*}^{SOP} + \lambda_{i\,*\,*}^{SOP} + \lambda_{*\,j\,*}^{SOP} + \lambda_{*\,*\,k}^{SOP} + \\
& \lambda_{i\,j\,*}^{SOP} + \lambda_{i\,*\,k}^{SOP} + \lambda_{*\,j\,k}^{SOP} + \lambda_{i\,j\,k}^{SOP}
\end{aligned}
\tag{5.1}
$$

for $i, j, k = 1, 2, 3$. In Equation 5.1, $\lambda_{*\,*\,*}^{SOP}$ is the overall effect; the parame-
ters $\lambda_{i\,*\,*}^{SOP}, \lambda_{*\,j\,*}^{SOP}$, and $\lambda_{*\,*\,k}^{SOP}$ represent the main effects of the categories of
the three variables; the parameters $\lambda_{i\,j\,*}^{SOP}, \lambda_{i\,*\,k}^{SOP}$, and $\lambda_{*\,j\,k}^{SOP}$ represent the
two-variables interaction effects; finally, the parameters $\lambda_{i\,j\,k}^{SOP}$ represent the
three-variables interaction effects.

The log-linear model defined by Equation 5.1 is not yet identified. For
example, adding a constant to the terms $\lambda_{i\,*\,*}^{SOP}$ and subtracting the same
constant from the terms $\lambda_{*\,j\,*}^{SOP}$ will leave the value of $\log \pi_{i\,j\,k}^{SOP}$ unchanged.
In order to obtain identified models we have used dummy coding for the
log-linear effects by setting all log-linear parameters with $i = 1$, $j = 1$, or
$k = 1$ equal to zero. These identification constraints imply that all log-
linear effects are expressed as deviations from the first response category of
a variable.

The model introduced in Equation 5.1 defines a linear structure on the
logarithms of the cell probabilities. It can be written in matrix notation as

$$
\log \boldsymbol{\pi}^{SOP} = \mathbf{X} \boldsymbol{\lambda}^{SOP}
$$

for some appropriately defined design matrix $\mathbf{X}$ that takes into account the
identifying constraints and that is of full column rank. The notation we
use assumes that if a scalar function (such as log or exp) is applied to a
vector, it is actually applied to each coordinate of the vector. Hence, if

$$
\boldsymbol{\pi} = \begin{pmatrix} \pi_1 \\ \pi_2 \\ \pi_3 \end{pmatrix}
$$

then for any scalar function f, we have

$$f(\boldsymbol{\pi}) = \begin{pmatrix} f(\pi_1) \\ f(\pi_2) \\ f(\pi_3) \end{pmatrix}.$$

In Equation 5.1 the vector $\boldsymbol{\lambda}^{SOP}$ contains the log-linear parameters in the appropriate order. Any log-linear model can be written in this matrix form for some design matrix $\mathbf{X}$. When we fit a log-linear model to a contingency table, we try to obtain estimates of the log-linear parameters that minimize a discrepancy function that quantifies the distance between the observed and the expected frequencies. Most often the maximum likelihood principle is used in this context.

Any log-linear model can also be specified by the constraints it imposes on the theoretical cell probabilities. Let $\mathbf{U}$ be a full column rank matrix such that the vector space spanned by the columns of $\mathbf{U}$ is the orthocomplement of the vector space spanned by the columns of $\mathbf{X}$. If $\mathcal{V}$ is the vector space spanned by the columns of $\mathbf{X}$, the columns of $\mathbf{U}$ span the vector space consisting of all vectors that are orthogonal to $\mathcal{V}$. Then, we have $\mathbf{w} = \mathbf{Xa}$ for some vector $\mathbf{a}$ if and only if $\mathbf{U'w} = 0$.

Any log-linear model can then also be characterized by the constraints it imposes on the elements of vector $\boldsymbol{\pi}$. If

$$\log \boldsymbol{\pi}^{SOP} = \mathbf{X}\boldsymbol{\lambda}^{SOP}$$

for some appropriately defined design matrix $\mathbf{X}$, then also

$$\mathbf{U'} \log \boldsymbol{\pi}^{SOP} = 0. \tag{5.2}$$

Instead of fitting a log-linear model by direct estimation of the log-linear parameters, we can fit it by determining the maximum likelihood estimates of the cell probabilities under the constraints given in Equation 5.2. As we will discuss later, this alternative way to characterize a log-linear model by the constraints it imposes on the cell probabilities easily generalizes to more complex models for contingency tables. For testing the goodness-of-fit of such complex models the approach that characterizes the model by its constraint equations is often the only feasible one because such models cannot be easily described in terms of parametric equations for the cell probabilities. This is especially the case for models that cannot be cast in a log-linear form.

The model defined by Equation 5.1 is the saturated model for Table 5.1. A saturated model always provides a perfect fit for the data at hand, and consequently, cannot be rejected on the basis of a statistical test. However, saturated models serve as baseline models against which the fit of more restrictive, parsimonious models can be tested. Each of the more restrictive models that are considered in the rest of this section can be characterized by the linear constraints it imposes on the logarithms of the cell probabilities for Table 5.1. Hence, each of these models can be represented by an equation like Equation 5.2 for an appropriately defined matrix **U**. Space limitations prohibit us to present the matrix **U** for each of the models we will consider.

A reasonable way to start a search for a more parsimonious model is to consider the model with no three-variables interaction terms. This model is derived from the saturated model by putting all the three-variables interaction terms equal to zero

$$\lambda_{i\,j\,k}^{SOP} = 0.$$

This model is often symbolically represented by $[SO, SP, OP]$ which is the list of highest-order interaction terms included in the model. For the present data, this model fits very well with the log likelihood ratio test $G^2 = 7.7988$, which for $df = 8$ gives $p = 0.453$.

In the search for a more parsimonious model that still provides a statistically acceptable fit, substantive arguments should play a decisive role. Since it may be expected that someone's opinion on CT depends on his political orientation, one could test the log-linear model $[SP, OP]$ which is derived from the previous model by omitting the interaction terms for the pair of variables S and O. This model implies that the association between the two variables S and O is completely spurious as a result of their association with the third variable P. However, this model does not fit the data with $G^2 = 27.1476$, which for $df = 12$ gives $p = 0.007$. The conditional test of the last model against the model $[SO, SP, OP]$ is even clearer: $G^2 = 19.3488$ giving $p = 0.0007$ with $df = 4$. These results lead to a rejection of the more restrictive model, and show that not all of the association between S and O is explained by P.

An alternative way to define a more parsimonious model is to assume that the log-linear structure of the association between Political Orientation and Opinion does not change from September to October. This hypothesis imposes the following equality constraints on some log-linear parameters in the $[SO, SP, OP]$ model

$$\lambda_{i\,*\,k}^{SOP} = \lambda_{*\,i\,k}^{SOP} \tag{5.3}$$

for $i = 1, 2, 3$ and $k = 1, 2, 3$. This more parsimonious model fits the data very well: The unconditional test yields $G^2 = 13.152$ with $df = 15$ and

Table 5.2: Marginal Table for Opinion in September Against Political Orientation.

		P	
S	1	2	3
1	29	115	123
2	54	60	23
3	129	276	138

$p = 0.591$. The conditional test of this model against the no-three-variables interaction model gives $G^2 = 5.354$ with $df = 7$ and $p = 0.617$. The fact that the parsimonious model defined by Equation 5.3 can be accepted for the present data implies that important aspects of the association between Political Orientation and Opinion do not change over time. The equality constraints on the log-linear terms in this model are equivalent to the following equality constraints on the cell probabilities

$$\frac{\pi_{ijk}^{SOP}}{\pi_{jik}^{SOP}} = \frac{\pi_{ij\ell}^{SOP}}{\pi_{ji\ell}^{SOP}}.$$

Hence, the odds of responding to the opinion question with the categories i and j in that temporal order rather than in the reverse order do not depend on Political Orientation.

5.3 Log-Linear Models for Marginal Tables

In this section we discuss how models for several marginal tables defined on the same original joint table can be tested simultaneously by means of a generalized log-linear analysis. We illustrate this approach by further analyses on the data from Table 5.1. More specifically, we investigate the stability of Opinion over time in its relation with Political Orientation by comparing the bivariate marginal distributions SP and OP. The observed frequencies of both marginal distributions are given in Table 5.2 and Table 5.3.

The cell probabilities in Tables 5.2 and 5.3 are denoted by π_{ik}^{SP} and by π_{ik}^{OP}, respectively. They are related to the cell probabilities π_{ijk}^{SOP} in Table 5.1 in the following way

$$\pi_{ik}^{SP} = \sum_{j} \pi_{ijk}^{SOP},$$

Table 5.3: Marginal Table for Opinion in October Against Political Orientation.

		P	
O	1	2	3
1	65	197	162
2	79	84	38
3	68	170	84

and

$$\pi_{i\ k}^{OP} = \sum_j \pi_{j\ i\ k}^{SOP}.$$

With all cell probabilities collected in the vectors π^{SP}, π^{OP} and π^{SOP}, we have for appropriately defined constant matrices $\mathbf{A}_1$ and $\mathbf{A}_2$

$$\left(\begin{array}{c} \pi^{SP} \\ \pi_{OP} \end{array} \right) = \left(\begin{array}{c} \mathbf{A}_1 \\ \mathbf{A}_2 \end{array} \right) \cdot \pi^{SOP}$$

In the present application, both $\mathbf{A}_1$ and $\mathbf{A}_2$ are matrices of order 9×27 whose elements are either 0 or 1. These matrices pick and add the appropriate elements of the vector π^{SOP} in order to obtain the marginal probabilities in the vectors π^{SP} and π^{OP}.

A first hypothesis that could be tested when comparing the two bivariate marginal distributions is whether these distributions are identical, or not. Hence, we would test whether

$$\pi_{i\ k}^{SP} = \pi_{i\ k}^{OP} \tag{5.4}$$

holds for all i and k. This is the hypothesis of bivariate marginal homogeneity which is here equivalent to the statement that the joint distribution of Political Orientation and Opinion on CT does not change over time. Since they are both derived from the three-dimensional table, the bivariate marginal tables given in Tables 5.2 and 5.3 are not statistically independent. Hence, equality of the marginal distributions cannot be tested by standard chi-square tests that assume independence of the two samples. To circumvent this problem, we rewrite the hypothesis of bivariate marginal homogeneity in the form of constraint equations on the cell probabilities $\pi_{i\ j\ k}^{SOP}$ that refer to the total table. The hypothesis of bivariate marginal homogeneity can then be tested by checking whether these constraints on the cell probabilities in the total table hold for the data at hand.

Define

$$\mathbf{A} = \left(\begin{array}{c} \mathbf{A}_1 \\ \mathbf{A}_2 \end{array} \right) \text{ and } \mathbf{U} = \left(\begin{array}{c} \mathbf{I}_{9 \times 9} \\ -\mathbf{I}_{9 \times 9} \end{array} \right).$$

The bivariate marginal homogeneity model described by Equation 5.4 is equivalent to the following constraints on the elements of π^{SOP}

$$\mathbf{U}'.\log(\mathbf{A}.\pi^{SOP}) = \mathbf{0}. \tag{5.5}$$

This matrix equation is very similar to the one for the log-linear model defined by Equation 5.1. For the latter model the matrix $\mathbf{A}$ is simply a 27×27 identity matrix, whereas for models defined on marginal tables, the matrix $\mathbf{A}$ selects and adds the appropriate elements from π^{SOP} that contribute to a particular marginal cell probability. All the models that are discussed further on in this section can be written in this form for appropriately defined matrices $\mathbf{A}$ and $\mathbf{U}$.

The model of bivariate marginal homogeneity as defined by Equation 5.4 fits the data very badly: It results in $G^2 = 138.091$ with $df = 8$ and $p = 0.000$. It is clear that the joint distribution of Opinion and Political Orientation changes over time.

A less stringent hypothesis about similarities between the two bivariate marginal distributions states that corresponding local odds ratios from both marginals are equal

$$\frac{\pi_{i\,j}^{SP} \pi_{i+1,j+1}^{SP}}{\pi_{i+1,j}^{SP} \pi_{i,j+1}^{SP}} = \frac{\pi_{i\,j}^{OP} \pi_{i+1,j+1}^{OP}}{\pi_{i+1,j}^{OP} \pi_{i,j+1}^{OP}} \tag{5.6}$$

for $i = 1, 2$ and $j = 1, 2$. This model is equivalent to a model that equates the two-variables interaction parameters in the log-linear models for the marginal tables. If these latter models are written as

$$\log \pi_{i\,k}^{SP} = \lambda_{**}^{SP} + \lambda_{i*}^{SP} + \lambda_{*k}^{SP} + \lambda_{i\,k}^{SP},$$

and

$$\log \pi_{i\,k}^{OP} = \lambda_{**}^{OP} + \lambda_{i+}^{OP} * + \lambda_{*k}^{OP} + \lambda_{i\,k}^{OP},$$

an analysis under the hypothesis of equal local odds ratios is equivalent to an analysis that imposes equality constraints of the type

$$\lambda_{i\,k}^{SP} = \lambda_{i\,k}^{OP} \tag{5.7}$$

on the parameters in the log-linear models for the marginal tables. Although these constraints are very similar to the constraints given by Equation 5.3, it should be noted that they are not equivalent: Whereas Equation 5.3 imposes constraints on parameters of the log-linear model for the

Table 5.4: Univariate Distributions of Opinion in September and October.

Opinion	Sept.	Oct.
1	267	424
2	137	201
3	543	322

total table, Equation 5.7 imposes equality constraints on parameters from the separate log-linear models for the marginal tables. In general we have

$$\lambda_{i\,k}^{SP} \neq \lambda_{i\,*\,k}^{SOP}$$

and

$$\lambda_{i\,k}^{OP} \neq \lambda_{*\,i\,k}^{SOP}.$$

The model defined by Equation 5.6 fits the data very well: $G^2 = 7.719$ with $df = 4$ and $p = 0.108$. The difference between the two bivariate marginal distributions is clearly not due to a change in the structure of the association between Opinion and Political Orientation but to other aspects of the distributions. As the following analyses show, the most striking difference between the two observed marginal tables is the number of respondents that have no clear opinion at one of the interviews. Although in September 543 (57.3%) respondents have no clear opinion, this number decreases to 322 (34.0%) in October. This observation suggests that we should compare the univariate distributions of the opinion variable at the two time points. Table 5.4 gives the observed frequency distributions for opinion in September and October. The hypothesis of univariate marginal homogeneity, defined by the constraints

$$\pi_i^S = \pi_i^O,$$

for $i = 1, 2, 3$ fits the data very badly: $G^2 = 134.986$ with $df = 2$ and $p = 0.000$.

Now consider the model that states that the proportion of respondents with a favorable opinion among those with a clear opinion on CT does not change over time. This model that imposes the single constraint

$$\frac{\pi_1^S}{\pi_1^S + \pi_2^S} = \frac{\pi_1^O}{\pi_1^O + \pi_2^O},$$

fits the data very well: $G^2 = 0.638$ with $df = 1$ and $p = 0.424$. The estimate of the constant proportion of respondents with a favorable opinion among those with a clear opinion on CT is 0.456. Hence, the difference between the two univariate marginals seems due to the fact that a significantly larger number of respondents express a definite opinion on CT in October than in September. The same change in the probability of stating a definite opinion is also responsible for the significant difference between the bivariate marginal distributions.

It is important to note here that all models tested in this section can be characterized by the linear constraints they impose on the logarithms of certain linear combinations of the cell probabilities in $\boldsymbol{\pi}^{SOP}$. Hence, each of these models is described by an equation like Equation 5.5 for some appropriately defined matrices $\mathbf{U}$ and $\mathbf{A}$.

5.4 Marginal Models That Cannot Be Cast Into a Log-Linear Form

In this section we discuss how models that cannot be rewritten as log-linear models on marginal tables can be fitted and tested by a generalization of the methods described in the previous sections. As an illustration we further study the association between Political Orientation and Opinion on CT and show how hypotheses about changes in the value of the association coefficient γ or the effect estimate ϵ can be tested by this approach. See Liebetrau (1983) and Gibbons (1993) for an elementary introduction to these and other association coefficients.

We first illustrate for the simple case of a 2×2 table how both coefficients can be represented by a recursively defined expression in terms of the cell probabilities. Suppose we have a 2×2 table for the cross-classification of two variables X and Y. Let the matrix with the theoretical cell probabilities in the 2×2 table be given by

$$\begin{pmatrix} \pi_{11} & \pi_{12} \\ \pi_{21} & \pi_{22} \end{pmatrix}$$

with rows corresponding to X and columns to Y. The same cell probabilities can be stacked row-wise in a column vector like

$$\boldsymbol{\pi} = \begin{pmatrix} \pi_{11} \\ \pi_{12} \\ \pi_{21} \\ \pi_{22} \end{pmatrix}.$$

For a 2×2 table, the coefficient γ is given by

$$\gamma = \frac{\pi_{11}\pi_{22} - \pi_{12}\pi_{21}}{\pi_{11}\pi_{22} + \pi_{12}\pi_{21}};$$

the ϵ-coefficient is defined by

$$\epsilon = \frac{\pi_{22}}{\pi_{21} + \pi_{22}} - \frac{\pi_{12}}{\pi_{11} + \pi_{12}}.$$

In computing the association coefficient γ, both variables are treated symmetrically, but in computing the coefficient ϵ, variable X is considered as the independent variable that has an effect on the dependent variable Y. Now define the following matrices

$$\mathbf{A}_1 = \mathbf{I}_{4 \times 4}$$

$$\mathbf{A}_2 = \begin{pmatrix} 1 & 0 & 0 & 1 \\ 0 & 1 & 1 & 0 \end{pmatrix}$$

$$\mathbf{A}_3 = \begin{pmatrix} 1 & 0 \\ 0 & 1 \\ 1 & 1 \end{pmatrix}$$

$$\mathbf{A}_4 = \begin{pmatrix} 1 & 0 & -1 \\ 0 & 1 & -1 \end{pmatrix}$$

and

$$\mathbf{A}_5 = \begin{pmatrix} 1 & -1 \end{pmatrix}.$$

A tedious but straightforward derivation shows that

$$\gamma = \mathbf{A}_5 . \exp(\mathbf{A}_4 . \log(\mathbf{A}_3 . \exp(\mathbf{A}_2 . \log(\mathbf{A}_1 . \pi)))).$$

The ϵ-coefficient can be represented in a similar way if we define

$$\mathbf{A}_1 = \begin{pmatrix} 0 & 1 & 0 & 0 \\ 1 & 1 & 0 & 0 \\ 0 & 0 & 0 & 1 \\ 0 & 0 & 1 & 1 \end{pmatrix}$$

$$\mathbf{A}_2 = \begin{pmatrix} 1 & -1 & 0 & 0 \\ 0 & 0 & 1 & -1 \end{pmatrix}$$

and

$$\mathbf{A}_3 = \begin{pmatrix} -1 & 1 \end{pmatrix}.$$

Then

$$\epsilon = \mathbf{A}_3 . \exp(\mathbf{A}_2 . \log(\mathbf{A}_1 . \pi)).$$

The expressions for γ and ϵ are both generalizations of the notation we have for specifying log-linear models. Both expressions can be written recursively in terms of a sequence of scalar functions f_k.

For the γ coefficient we need five functions defined as follows:

$$\begin{aligned}
f_1(x) &= f_3(x) = \log(x) \\
f_2(x) &= f_4(x) = \exp(x) \\
f_5(x) &= x
\end{aligned}$$

Now let $\mathbf{u}_0 = \pi$ and define recursively

$$\mathbf{u}_k = f_k(\mathbf{A}_k \mathbf{u}_{k-1}).$$

Then it follows that

$$\begin{aligned}
\gamma &= \mathbf{u}_5 \\
&= f_5(\mathbf{A}_5 . \mathbf{u}_4) \\
&= f_5(\mathbf{A}_5 . f_4(\mathbf{A}_4 . \mathbf{u}_3)) \\
&= f_5(\mathbf{A}_5 . f_4(\mathbf{A}_4 . f_3(\mathbf{A}_3 . \mathbf{u}_2))) \\
&= f_5(\mathbf{A}_5 . f_4(\mathbf{A}_4 . f_3(\mathbf{A}_3 . f_2(\mathbf{A}_2 . \mathbf{u}_1)))) \\
&= f_5(\mathbf{A}_5 . f_4(\mathbf{A}_4 . f_3(\mathbf{A}_3 . f_2(\mathbf{A}_2 . f_1(\mathbf{A}_1 . \mathbf{u}_0))))).
\end{aligned}$$

For the ϵ coefficient we need three functions defined as follows

$$\begin{aligned}
f_1(x) &= \log(x) \\
f_2(x) &= \exp(x) \\
f_3(x) &= x
\end{aligned}$$

Then

$$\begin{aligned}
\epsilon &= \mathbf{u}_3 \\
&= f_3(\mathbf{A}_3 . \mathbf{u}_2) \\
&= f_3(\mathbf{A}_3 . f_2(\mathbf{A}_2 . \mathbf{u}_1)) \\
&= f_3(\mathbf{A}_3 . f_2(\mathbf{A}_2 . f_1(\mathbf{A}_1 . \mathbf{u}_0))).
\end{aligned}$$

Both γ and ϵ can be defined for general $R \times C$ contingency tables. Let π_{ij} denote an arbitrary cell probability from this table. Now suppose we have drawn two observations from this table. These observations are concordant if one of them scores higher on both variables. They are discordant if one of them scores higher on one variable and lower on the other variables. A tie occurs if both observations score the same on at least one variable. Coefficient γ is then given by

$$\gamma = \frac{\Pi_{con} - \Pi_{dis}}{\Pi_{con} + \Pi_{dis}},$$

with the probabilities of concordance and discordance defined as

$$\Pi_{con} = \sum_{i=1}^{R-1} \sum_{j=1}^{C-1} \left(\pi_{ij} \sum_{k=i+1}^{R} \sum_{\ell=j+1}^{C} \pi_{k\ell} \right),$$

and

$$\Pi_{dis} = \sum_{i=1}^{R-1} \sum_{j=2}^{C} \left(\pi_{ij} \sum_{k=i+1}^{R} \sum_{\ell=1}^{j-1} \pi_{k\ell} \right).$$

Note that γ is an association coefficient that assumes that the response categories of both variables are ordered.

The generalization of ϵ to arbitrary $R \times C$ tables is somewhat more involved since $(R-1)(C-1)$ linearly independent different ϵ values can be obtained. For $i = 1, \cdots, R-1$ and $j = 1, \cdots, C-1$, one may define

$$\epsilon_{ij} = \pi_{j|i+1} - \pi_{j|i}.$$

Note that in defining the ϵ coefficients, one of the variables has to be chosen as the response variable.

The recursive definitions of γ and ϵ given above for a 2×2 table can easily be generalized to arbitrary $R \times C$ contingency tables by extending the A-matrices in appropriate ways. Moreover, hypotheses about equality of γ's or ϵ's for the same variables measured in different subgroups, or for the same variables measured at different time points can easily be formulated within this approach but requires in general a few more steps in the recursive definition. As an illustration we investigate whether the association between Political Orientation and Opinion changes over time.

We first test the hypothesis that γ remains constant over time. Because it is not clear how for the opinion variables the response category No clear opinion can be meaningfully ordered relative to the two other categories, we decided to retain in the computation of γ only those respondents with a clear opinion at the time of measurement. So, γ_{SP} between Opinion and Political Orientation is computed on a 2×3 table that corresponds to the

first two rows of Table 5.2. Similarly, γ_{OP} is computed on a similarly selected subtable from Table 5.3. In both computations the response category
Favorable is taken as the positive response, the category Not Favorable as
the negative response. Note that this way of selecting the subjects has the
effect that the two γs are computed on partially different samples: γ_{SP} is
based on the sample of 404 respondents with a definite opinion in September, γ_{OP} is based on the sample of 625 respondents with a definite opinion
in October, and only the 322 respondents who had a clear opinion at both
measurement points contributed to both coefficients.

The estimates of the coefficients are $\gamma_{SP} = 0.600$ and $\gamma_{OP} = 0.463$. The
test of the hypothesis that both γ's are equal yields $G^2 = 4.02$ with $df = 1$
and $p = 0.045$. There seems to be some evidence that the strength of the
association between Opinion and Political Orientation is somewhat smaller
in October than in September, but the difference is not outspoken. At both
time points it is clear that conservative respondents have a more favorable
opinion than liberal respondents.

Next we test whether the same conclusions can be drawn in terms of
the ϵ coefficient. In the present example it seems reasonable to assume
that Political Orientation has an effect on Opinion, and not the other way
around. Hence, Opinion is taken as the response variable in the present
application. Since Political Orientation has three response categories, two
different linearly independent ϵ should be considered: ϵ_{21} which is the difference in the probability of a Favorable opinion between Moderates versus
Liberals, and ϵ_{32}, which is the difference in the probability of a Favorable
opinion between Conservatives versus Moderates.

The observed values of these two ϵ coefficients are $\epsilon_{21} = 0.308$ and
$\epsilon_{32} = 0.185$ in September, and $\epsilon_{21} = 0.250$ and $\epsilon_{32} = 0.109$ in October. Note
that here too the ϵs of September and October are computed on samples
of respondents that only partially overlap. Corresponding values of the ϵs
are somewhat lower in October than in September, a result that seems to
confirm our previous conclusion based on γ. However, the hypothesis that
corresponding ϵ values are equal, that is

$$\epsilon_{21}^{SP} = \epsilon_{21}^{OP}$$

and

$$\epsilon_{32}^{SP} = \epsilon_{32}^{OP}$$

cannot be rejected since $G^2 = 5.475$ with $df = 2$ and $p = 0.065$. Here,
the conclusion is that the strength of the effect of Political Orientation on
Opinion does not change over time. The estimates of the constant value of
the ϵs under this hypothesis are $\epsilon_{21} = 0.266$ and $\epsilon_{32} = 0.134$. The size of

the ϵ's and their signs indicate that the more conservative one's political orientation, the more favorable one's opinion on CT.

5.5   Estimating and Testing Models Specified by Constraints on the Cell Probabilities

In this section we discuss the more technical details of the optimization algorithm for estimating and testing a model that is specified by constraints on the cell probabilities.

Suppose that in a particular investigation K categorical variables are measured on a sample of N respondents. If m_k represents the number of response categories for variable k, the number of different response patterns is given by $M = \prod_k m_k$. If the response patterns are numbered in a particular way from 1 to M, we let π_i represent the probability of observing response pattern i in the sample and $\pi' = (\pi_1, \cdots, \pi_M)$ the vector with all these response pattern probabilities. The vector $\mathbf{f}' = (f_1, \cdots, f_M)$ contains the corresponding observed frequencies.

All the models discussed in the previous sections can be specified by a set of constraint equations on the cell probabilities in the original total table. Let these constraint equations be collected in vector

$$\mathbf{h}(\boldsymbol{\pi}) = \begin{pmatrix} h_1(\boldsymbol{\pi}) \\ \vdots \\ h_R(\boldsymbol{\pi}) \end{pmatrix} = \mathbf{0}_{r \times 1},$$

with each constraint recursively defined in terms of appropriate scalar functions and matrices.

Assuming multinomial sampling, the maximum likelihood estimates of the cell probabilities are obtained by means of an iterative procedure that determines a saddlepoint of the Lagrangian

$$\mathbf{f}'.\log(\boldsymbol{\pi}) - \boldsymbol{\mu}'\mathbf{h}(\boldsymbol{\pi}),$$

in which $\boldsymbol{\mu}$ is a vector containing R unknown Lagrangian multipliers.

Next, we describe a Fisher scoring algorithm proposed by Bergsma (1997) to find the MLE. Let $\mathbf{m} = N\boldsymbol{\pi}$ be the expected cell frequencies under the model that is being tested and define the augmented likelihood

$$L(\mathbf{m}, \boldsymbol{\mu}) = \mathbf{f}'.\log(\mathbf{m}) - \boldsymbol{\mu}'\mathbf{h}(\mathbf{m}).$$

Let $\mathbf{H} = \mathbf{H}(\mathbf{m})$ be the Jacobian of $\mathbf{h}(\mathbf{m})$ with respect to $\log \mathbf{m}$. Then differentiating $L(\mathbf{m}, \boldsymbol{\mu})$ with respect to $\log \mathbf{m}$ yields

$$\mathbf{l}(\mathbf{m}, \boldsymbol{\mu}) = \mathbf{f} - \mathbf{m} - \mathbf{H}\boldsymbol{\mu}.$$

Under suitable regularity conditions, the MLE $\hat{m}$ is a vector m for which there is a Lagrange multiplier vector μ such that the simultaneous equations

$$\begin{aligned} l(m, \mu) &= 0 \\ h(m) &= 0 \end{aligned}$$

are satisfied.

Let $D(m)$ be a diagonal matrix with the vector m on its main diagonal. The expected value of the derivative matrix of the vector $(l(m, \mu), h(m))$ with respect to (m, μ) then is

$$V(m) = \begin{pmatrix} -D(m) & H \\ H' & 0 \end{pmatrix}.$$

With f^+ equal to the vector f with zeroes replaced by a small positive constant (say, 10^{-10}), define the Fisher scoring starting values

$$\begin{pmatrix} \log m^{(0)} \\ \mu^{(0)} \end{pmatrix} = \begin{pmatrix} \log f^+ \\ 0 \end{pmatrix}$$

and, for $k = 0, 1, \ldots$

$$\begin{pmatrix} \log m^{(k+1)} \\ \mu^{(k+1)} \end{pmatrix} = \begin{pmatrix} \log m^{(k)} \\ \mu^{(k)} \end{pmatrix} - V(m^{(k)})^{-1} \cdot \begin{pmatrix} l(m^{(k)}, \mu^{(k)}) \\ h(m^{(k)}) \end{pmatrix}.$$

Then as $k \to \infty$, $m^{(k)}$ should go to $\hat{m}$. Straightforward matrix algebra yields the simplified form

$$\begin{aligned} \log m^{(k+1)} &= \log m^{(k)} + D\left(m^{(k)}\right)^{-1} l\left(m^{(k)}, \mu^{(k+1)}\right) \\ \mu^{(k+1)} &= -\left(H'D\left(m^{(k)}\right)H\right)^{-1} \left(H'D\left(m^{(k)}\right)^{-1} \left(f - m^{(k)}\right)\right. \\ &\quad \left. + h\left(m^{(k)}\right)\right) \end{aligned}$$

(see Bergsma, 1997). This algorithm does not always converge, and it can be helpful to introduce a step size $step^{(k)} \in (0, 1]$ as follows

$$\log m^{(k+1)} = \log m^{(k)} + step^{(k)} D\left(m^{(k)}\right)^{-1} l\left(m^{(k)}, \mu^{(k+1)}\right)$$

Note that the update of μ is left unchanged.

The step size should be chosen so that the new estimate $m^{(k+1)}$ is "better" than the old estimate $m^{(k)}$. A criterion for deciding this is obtained by defining the following quadratic form measuring the "distance" from convergence

$$d\left(\mathbf{m}^{(k)}\right) = 1\left(\mathbf{m}^{(k)}, \boldsymbol{\mu}^{(k+1)}\right)' \mathbf{D}\left(\mathbf{m}^{(k)}\right)^{-1} 1\left(\mathbf{m}^{(k)}, \boldsymbol{\mu}^{(k+1)}\right).$$

Convergence is reached at $\mathbf{m}$ if and only if $d(\mathbf{m}) = 0$, and therefore, if possible, the step size should be chosen so that $d(\mathbf{m}^{(k+1)}) < d(\mathbf{m}^{(k)})$ for all k. Sufficiently close to the MLE this is possible, but otherwise not necessarily. In such cases, a recommendation which seems to work very well in practice is to "jump" to another region by taking a step size equal to one.

After convergence of the estimation procedure, the null hypothesis that the model specified by the R constraints $\mathbf{h}(\boldsymbol{\pi}) = \mathbf{0}$ provides an acceptable fit to the data can be tested against the saturated model by means of a log likelihood ratio test. The test statistic is

$$G^2 = 2\mathbf{f}'.\log(\mathbf{f}/\hat{\mathbf{m}}),$$

which follows asymptotically a chi square distribution with R degrees of freedom if the null hypothesis is true.

A *Mathematica* source code for the maximum likelihood fitting of models specified by constraint equations on the cell probabilities is available from the authors.

5.6   Discussion

The examples given above illustrate the versatility of the approach in which marginal models are characterized by the constraints they impose on the cell probabilities. Many models for a multidimensional contingency table or for some of its marginal tables can be specified by the nonlinear constraints they impose on the theoretical cell probabilities of the original table. By using a flexible recursive notation that involves sequences of constant matrices and scalar functions, these nonlinear constraints can be represented in such a way that they can easily be incorporated in a constrained maximum likelihood estimation procedure for the cell probabilities. An advantage of this approach is that the models need not be written explicitly in terms of model parameters.

In the present chapter we have illustrated this general approach to marginal modelling by discussing how it can be used to test hypotheses about association coefficients such as γ and ϵ. However, many more hypotheses about the distributions of the variables involved may be tested in this way. Among these, hypotheses about differences among coefficients for central tendency and variability are the most prominent. Also hypotheses in terms of information-theoretic measures can be tackled by this approach. In a forthcoming book (Bergsma, Croon, & Hagenaars, 2004) many types of marginal models will be considered and treated from the point of view that was advocated in this chapter.

References

Aitchison, J. (1962). Large-sample restricted parametric tests. *Journal of the Royal Statistical Society, Series B, 24*, 234-250.

Aitchison, J., & Silvey, S. D. (1958). Maximum likelihood estimation of parameters subject to constraints. *Annals of Mathematical Statistics, 29*, 813-828.

Aitchison, J., & Silvey, S. D. (1960). Maximum likelihood procedures and associated tests of significance. *Journal of the Royal Statistical Society, Series B, 22*, 154-171.

Bergsma, W. P. (1997). *Marginal models for categorical data.* Tilburg: Tilburg University Press.

Bergsma, W. P., Croon, M., & Hagenaars, J. A. (2004). *Marginal models for categorical data.* Manuscript in preparation.

Croon, M. A., Bergsma, W. P., & Hagenaars, J. A. (2000). Analyzing change in categorical variables by generalized log-linear models. *Sociological Methods and Research, 29*, 195-229.

Forthofer, R. N., & Koch, G. G. (1973). An analysis for compounded functions of categorical data. *Biometrics, 29*, 143-157.

Gibbons, J. D. (1993). *Nonparametric measures of association.* Thousand Oaks, CA: Sage.

Grizzle, J. E., Starmer, C. F., & Koch, G. G. (1969). Analysis of categorical data by linear models. *Biometrics, 25*, 489-504.

Kritzer, H. M. (1977). Analyzing measures of association derived from contingency tables. *Sociological Methods and Research, 5*, 35-50.

Lang, J. B. (1996). Maximum likelihood methods for a generalized class of log-linear models. *The Annals of Statistics, 24*, 726-752.

Liebetrau, A. M. (1983). *Measures of association.* Thousand Oaks, CA: Sage.

Vermunt, J. K., Rodrigo, M. F., & Ato-Garcia, M. (2001). Modeling joint and marginal distributions in the analysis of categorical panel data. *Sociological Methods and Research, 30*, 170-196.

Chapter 6

Computational Aspects of the E-M and Bayesian Estimation in Latent Variable Models

Irini Moustaki
Athens University of Economics and Business

Martin Knott
London School of Economics and Political Science

6.1 Introduction

Latent variable models are widely used in Social Sciences for measuring constructs (latent variables) such as ability, belief, attitude, behavior, welfare, satisfaction, and knowledge. Those unobserved constructs are measured through a number of observed indicators (items). Those indicators may be binary categorical (yes/no), ordinal categorical (strongly disagree, disagree, agree, strongly agree), nominal (political party, religion), or metric (weight, height, income, expenditures). Latent variable models have two main objectives. The first objective is to reduce the dimensionality of multivariate observed data by trying to identify a small number of latent dimensions that can explain the relationships among the observed items. The second objective is to score population members on those identified latent dimensions based on what they have responded to the observed items. Those latent scores can sometimes be used in regression models as dependent or independent variables. However, latent variable models have also been recently extended to allow effects of explanatory variables directly on the latent variables and/or on the indicator variables (Moustaki, 2003; Sammel, Ryan, & Legler, 1997). One can distinguish two main groups of estimation methods for latent variable models. The first group of methods is based on Monte Carlo Markov Chain Bayes estimation (MCMC). The second group includes methods based either on Newton-Raphson iteration or on the E-M algorithm. MCMC methods have recently become popular in the area of latent variable modelling mainly because they allow estimation of complex models (see e.g., Albert, 1992; Albert & Chib, 1993; Patz & Junker, 1999a, 1999b; Dunson, 2000). In the psychometric, educational and medical literature models have been developed for binary, nominal, ordinal, and metric manifest indicators see, for example, Bartholomew and Knott (1999), Bock and Aitkin (1981), Bock, Gibbons, and Muraki (1988), Muraki and Carlson (1995), and Samejima (1969). Moustaki (1996) and Sammel et al. (1997) also looked at models with mixed metric and binary observed indicators and recently Moustaki (2000a) and Moustaki and Knott (2000a) presented a general exponential family framework for fitting latent variable models to any type of observed items or to data including several different types of items. The models discussed in those papers are estimated by marginal maximum likelihood using an E-M algorithm.

The aim of the chapter is to compare the E-M and the MCMC estimation methods for latent variable models. The comparison is made in terms of parameter estimates, standard errors and computation time needed. Real examples with categorical indicators are used to compare the E-M approach to the MCMC method. For the comparison we use the software GENLAT (Moustaki, 2000b) for the E-M method. GENLAT has been written in FORTRAN and it gives parameter estimates, standard errors, factor scores and goodness-of-fit statistics. The program deals with missing values as well. Examples of other commercial software using estimation methods similar to GENLAT are the programs BILOG (Mislevy & Bock, 1990) and TESTFACT (Wilson, Wood, & Gibbons, 1991) for binary responses and the program MULTILOG (Thissen, 1991) for ordinal responses. For the MCMC estimation we use the software BUGS (Bayesian inference Using Gibbs Sampling, Spiegelhalter, Thomas, Best, & Gilks, 1996), which can be downloaded for free from the url *http://www.mrc-bsu.cam.ac.uk/bugs/*.

6.2 Latent Variable Models for Binary and Ordinal Responses

Let us denote by $(y_1, \ldots, y_p)$ the p observed variables (we often call them *indicators*) that may in principle be of any type, but here will be either all binary or all ordinal responses.

6.2.1 Binary Responses

Let us assume that the observed indicator y_i takes values 0 and 1.

For binary indicators we fit the logistic model to the probability of getting a positive/correct response 1 conditional on the latent variables, denoted by $\mathbf{z} = (z_1, \ldots, z_q)$. The model is

$$\text{logit } \pi_i(\mathbf{z}) = \alpha_{i0} + \sum_{j=1}^{q} \alpha_{ij} z_j$$

where $\pi_i(\mathbf{z}) = P(y_i = 1 \mid \mathbf{z})$.

The conditional probability of the variable y_i given the vector of latent variables $\mathbf{z}$ is

$$g(y_i \mid \mathbf{z}) = \pi_i(\mathbf{z})^{y_i} (1 - \pi_i(\mathbf{z}))^{1-y_i}.$$

The parameter α_{i0} is used to compute the probability that the median individual ($\mathbf{z} = 0$) will respond positively to item i from

$$\pi_i(0) = \frac{e^{\alpha_{i0}}}{1 + e^{\alpha_{i0}}};$$

the factor loadings α_{ij} show the effect of the latent variables on the logit of the probability of a correct response. For the case of one latent variable the model is known as the two parameter logistic model (Bock & Aitkin, 1981).

6.2.2 Ordinal Responses

If the observed variable y_i is a Likert type variable with C_i response categories the model often used in applications is the proportional odds model (see Agresti, 1990, pp. 322-331). In this we model the log of the odds of the cumulative probability, $\gamma_{is}(\mathbf{z})$, of being at or below a response category s as a function of the latent variables $\mathbf{z}$. The model is

$$\text{logit} \gamma_{is}(\mathbf{z}) = \alpha_{i0(s)} - \sum_{j=1}^{q} \alpha_{ij} z_j$$

where $\gamma_{is}(\mathbf{z}) = P(y_i \leq s \mid \mathbf{z})$ and $s = 1, \cdots, C_i - 1$. Note that $\gamma_{i,C_i} = 1$ by definition.

The conditional probability of the variable y_i given the vector of latent variables $\mathbf{z}$ is

$$g(y_i \mid \mathbf{z}) = \prod_{s=1}^{C_i} (\pi_i(\mathbf{z}))^{y_i(s)}$$

where $y_i(s)$ takes the value 1 if indicator i has category score s and 0 otherwise.

The parameters $\alpha_{i0(s)}$ are usually called cut-off points or thresholds and the α_{ij} are the factor loadings. The above model is known as the *proportional odds model* because it assumes that the effect of the latent variables is identical for all $C_i - 1$ collapsings of the responses into binary outcomes. For a unidimensional latent trait model the proportional odds model reduces to the graded response model introduced by Samejima (1969).

6.2.3   Model for Missing Values

Finally, we look at a model for binary responses with missing values as discussed in Knott, Albanese, & Galbraith (1990), O'Muircheartaigh and Moustaki (1999), Moustaki and Knott (2000b), and Moustaki and O'Muircheartaigh (2002).

To make the exposition of the model clearer, we present the results for a two dimensional factor model ($\mathbf{z} = (z_a, z_r)$) in which factor z_a represents the attitude dimension and z_r represents response propensity.

Suppose that we have p observed items (here of binary type) denoted by $y_1, \ldots, y_p$ for each of which a proportion of the individuals fail to respond. To reflect that lack of response we create p binary response propensity variables denoted by w_i as follows: when an individual gives a response to item i the response variable for this individual takes the value 1 ($w_i = 1$); when an individual does not give a response the response variable takes the value 0 ($w_i = 0$). The responses and nonresponses to the items are assumed to be independent conditional on the two latent variables z_a and z_r.

For each attitude binary item

$$Pr(y_i = 1 \mid z_a, z_r, w_i = 1) = \pi_{ai}(z_a).$$

For each response (pseudo) item:

$$Pr(w_i = 1 \mid z_a, z_r) = \pi_{ri}(z_a, z_r).$$

The logistic model already discussed above for binary responses becomes

$$\text{logit}\pi_{ai}(z_a) = \alpha_{i0} + \alpha_{i1} z_a \qquad (6.1)$$

and

$$\text{logit}\,\pi_{ri}(z_a, z_r) = r_{i0} + r_{i1}z_a + r_{i2}z_r.$$

The model allows attitude and response propensity to affect the probability of responding to an item leading to a model for nonignorable nonresponse. The model for binary responses is fitted to the $2 \times p$ indicators where the first p are attitudinal indicators and the last p are response propensity indicators. The factor loadings for the response propensity factor are fixed to zero for the p attitudinal items (see Equation 6.1).

The models presented above are all special cases of the generalized linear latent variable models framework and can all be fitted with the same likelihood (see Moustaki, 2000a; Moustaki & Knott, 2000a).

6.2.4 Maximum Likelihood Estimation and the E-M Algorithm

The number of parameters to be estimated depends on the type of the observed items. For the case of binary items we need to estimate α_{i0} and α_{ij} for all the items ($i = 1, \ldots, p$, $j = 1, \ldots, q$). In the ordinal case we need to estimate the thresholds $\alpha_{i0(s)}$ where $s = 1, \ldots, C_i - 1$ and the factor loadings α_{ij} where $i = 1, \ldots, p$, $j = 1, \ldots, q$.

We estimate the parameters by maximizing the marginal likelihood of the manifest variables $\mathbf{y}$. For a random sample of size n the complete loglikelihood of both manifest and latent variables $\mathbf{y}, \mathbf{z}$ can be written

$$\log f(\mathbf{y}, \mathbf{z}) = \sum_{m=1}^{n} \left[\sum_{i=1}^{p} \log g(y_{im} \mid \mathbf{z}_m) + \log h(\mathbf{z}_m) \right].$$

The latent variables $\mathbf{z}$ are taken to be independent with standard normal distributions so that $h(\mathbf{z})$ denotes a product of independent standard normal densities, and does not involve any model parameters. The model assumes conditional independence of the responses $\mathbf{y}$ given the vector $\mathbf{z}$ of latent variables. The conditional distribution of $g(y_{im} \mid \mathbf{z}_m)$ is taken here to be the Bernoulli and the multinomial for the binary and the ordinal items, respectively. The marginal likelihood is obtained by taking the expected value of the likelihood over the distribution of the latent variables.

The marginal likelihood solution is obtained using an E-M algorithm. Details can be found in Moustaki and Knott (2000a) and Moustaki (2000a). The routines GLLAMMA (Rabe-Hesketh, Pickles, & Skrondal, 2001), which run under STATA perform the same maximization using a Newton-Raphson algorithm.

The E-M algorithm in GENLAT requires initial values which can either be set by the program or read from a file. It is recommended to use different sets of initial values for each estimation as a check that the convergence of the algorithm is to a global rather than a local maximum. The convergence of the E-M algorithm is judged by the difference between successive likelihoods.

GENLAT computes asymptotic standard errors of the parameter estimates based on an approximation of the inverse of the information matrix at the maximum likelihood solution given by

$$I(\hat{\mathbf{a}}) = \left[\sum_{m=1}^{n} \frac{1}{f^2(\mathbf{y}_m, \mathbf{z}_m)} \frac{\partial f(\mathbf{y}_m, \mathbf{z}_m)}{\partial \mathbf{a}_j} \frac{\partial f(\mathbf{y}_m, \mathbf{z}_m)}{\partial \mathbf{a}_k} \right]^{-1}$$

where $\mathbf{a}$ is the vector of the estimated parameters.

Alternatively, the information matrix can be obtained within the E-M algorithm, see Louis (1982).

Instructions on how to use the program GENLAT are given on the URL address: *http://stats.lse.ac.uk/knott/software/latbug/*.

6.2.5 Bayesian Estimation Using BUGS

An alternative way of estimating the unknown parameters α_{i0}, $\alpha_{i0(s)}$ and α_{ij} is to use a Bayesian estimation approach. In the Bayesian approach the model parameters are not considered fixed but stochastic (random variables), and so probability statements are made about those parameters. In BUGS the model is defined as the joint distribution over all unobserved (parameters) and observed quantities (data). The way information is obtained about the unobserved quantities is through the posterior distribution of the unobserved quantities conditional on the data. The marginalization of the posterior distribution with respect to individual parameters of interest is done through Gibbs sampling.

Likelihood

Let us denote by $\mathbf{v}$ the vector with all the unknown parameters, $\mathbf{v}' = (\alpha_{i0}, \alpha_{i0(s)}, \alpha_{ij}, \mathbf{z})$. The log-likelihood is written as

$$L(\mathbf{y} \mid \mathbf{v}) = \sum_{h=1}^{n} \log f(\mathbf{y}_h) = \sum_{h=1}^{n} \log \int \cdots \int g(\mathbf{y}_h \mid \mathbf{v}) h(\mathbf{v}) d\mathbf{v}.$$

The posterior distribution of the parameter vector $\mathbf{v}$ is

$$h(\mathbf{v} \mid \mathbf{y}) = \frac{g(\mathbf{y} \mid \mathbf{v}) h(\mathbf{v})}{f(\mathbf{y})} \propto g(\mathbf{y} \mid \mathbf{v}) h(\mathbf{v}).$$

The form of the $g(\mathbf{y} \mid \mathbf{v})$ is from the exponential distribution with canonical parameter which is, for example, in the binary case equal to $\alpha_{i0} + \sum_{j=1}^{q} \alpha_{ij} z_j$. The main steps of the Bayesian approach are

1. Inference is based on the posterior distribution of the unknown parameters, α_{i0}, $\alpha_{i0(s)}$ α_{ij} and $\mathbf{z}$ conditional on what is known, here $\mathbf{y}$, $(h(\mathbf{v} \mid \mathbf{y}))$. Depending on the model fitted, the form of the posterior distribution can be very complex.

2. The mean vector of the posterior distribution $h(\mathbf{v} \mid \mathbf{y})$ can be used as an estimator of $\mathbf{v}$.

3. Standard errors for the parameter estimates can be computed from the standard deviation of $h(\mathbf{v} \mid \mathbf{y})$.

4. In general we may use the posterior mean $E(\psi(\mathbf{v})|\mathbf{y})$ as a point estimate of a function of the parameters $\psi(\mathbf{v})$, where $E(\psi(\mathbf{v}) \mid \mathbf{y}) = \int \cdots \int \psi(\mathbf{v}) h(\mathbf{v} \mid \mathbf{y}) d\mathbf{v}$.

5. Analytic evaluation of the above expectation is impossible. Alternatives include numerical evaluation, analytic approximations and Monte Carlo integration.

Markov Chain Monte Carlo, MCMC

To avoid the integration required in the posterior expectation, Monte Carlo integration is used that approximates the integrals by an average of quantities calculated from sampling. Samples are drawn from the posterior distribution of all the unknown parameters: $h(\mathbf{v}^{(i)} \mid \mathbf{y})$. Then the posterior expectation becomes

$$E(h(\mathbf{v} \mid \mathbf{y})) = \frac{1}{N} \sum_{i=1}^{N} h(\mathbf{v}^{(i)} \mid \mathbf{y}),$$

where N is the number of samples drawn.

The samples drawn from the posterior distribution do not have to be independent. Samples are drawn from the posterior distribution through a Markov chain with $h(\mathbf{v} \mid \mathbf{y})$ as its stationary distribution.

Algorithms such as the Gibbs sampler and Metropolis-Hastings are used in BUGS to get the unique stationary distribution.

Choosing prior distributions and starting values

For all the parameters defined in $\mathbf{v}$ we have to assume prior distributions. Diffuse or vague priors are assumed in all our applications so that the likelihood is emphasized rather than the prior. For example normal distributions with mean 0 and variance 100000 are assumed for all the α_{i0}, $\alpha_{i0(s)}$, and α_{ij} parameters. The latent variables $\mathbf{z}$ are assumed to have independent standard normal distributions.

BUGS can give to the random components of a model starting values itself but we found that the program works more efficiently if starting values are chosen by the user.

Convergence diagnostics in MCMC

A main concern in MCMC estimation methods is to assess at what stage the distribution of the parameter values produced by the Markov chain may be considered to be from the stationary distribution of the chain, which is the posterior distribution of the parameters given the data. Different criteria have been developed to check convergence. In BUGS there is a 'burn-in' period usually chosen by the user to be 500 or 1000 iterations before monitoring of the parameters through graphs and summary statistics starts.

The convergence criteria used in this chapter have been obtained from the program CODA (Best, Cowles, & Vines, 1996) which can also be obtained (manual and software) from the BUGS web address. The software CODA reads the BUGS output and produces a series of convergence diagnostics together with diagnostic plots. The plots that can be used to check convergence of each model parameter are the trace and the autocorrelation function. When we look at the autocorrelation function plot we would expect to see the autocorrelation reduces as the number of iteration increases. High autocorrelations within a chain are indicators of slow convergence. It is suggested in the CODA manual that a different parameterization of the model might reduce the autocorrelations. With respect to the trace plot (estimated values plotted against the iteration number presented as a time series plot) we expect that as the series converges the variability from one iteration to the other will decrease.

The convergence diagnostics are based on statistics developed by Geweke (1992) and Raftery and Lewis (1992b). Geweke's criterion uses two sections of the Markov chain. The first section contains the first 10% of the iterations and the second section contains the last 50% of the iterations. Those are the default values in CODA. If the whole chain is stationary then the mean of the parameter values in the early stage and in the late stage should be similar. The test is based on a z-score that compares the difference in the two means divided by the asymptotic standard error of the difference.

As the chain length tends to infinity the distribution of z tends to be a standard normal. Absolute values of z greater than 2 indicate that the chain has not converged. In the case of non convergence when the first 10% is compared with the last 50% one can disregard the first 10% and recompute the Geweke's diagnostic for comparing the first 20% with the last 50%.

The Raftery and Lewis (1992b) convergence criterion also applies to single chains. Their criterion aims to check the accuracy of the estimation of quantiles. The test computes the number of iterations required for achieving a certain level of accuracy in estimating quantiles. The default values in CODA for the quantile, the accuracy level and the probability are 2.5%, $+/-0.005$ and 0.95 respectively. A statistic computed within that criterion is that of the dependence factor that is computed as $I = N/Nmin$, where N is the maximum number of iterations needed to achieve convergence and $Nmin$ is the minimum number. The dependence factor measures the increase in the number of iterations needed to reach convergence due to dependence between the samples in the chain. Values of I greater than 1.0 indicate high within-chain correlations and probable convergence failure (see CODA manual). Raftery and Lewis (1992a) suggest that $I \geq 5.0$ indicates that a reparameterization of the model might be needed.

6.3 Examples

We next compare the results obtained from the two estimation methods in terms of parameter estimates, standard errors, estimation time, and diagnostic procedures available. We use three examples. In the url address *http://stats.lse.ac.uk/knott/software/latbug* we give the BUGS code used to estimate the different latent variable models. BUGS also provides a graphical interface (DOODLEBUGS) where users can specify their models using a graph.

The examples we chose vary in terms of sample size, number of items, and types of responses. The third data set includes missing values.

6.3.1 Latent Trait Model for Ordinal Responses

The first example is for ordinal responses. The data set consists of six ordinal items on environmental issues each with three response categories. The sample size is 291 respondents. The data set has been analyzed in Knott and Albanese (1993). For this chapter we have excluded responses with missing values and fitted the proportional odds model to the six ordinal items.

Table 6.1 gives the Geweke (1992) convergence diagnostic and the dependence factor I introduced by Raftery and Lewis (1992b) for the one-factor

Table 6.1: Geweke Convergence Diagnostic (z-score) and the Dependence Factor I for the One-Factor Model, Environment Data.

	Geweke criterion z-value	Dependence factor (I) for 2.5%	Dependence factor (I) for 50%
$\alpha_{10(1)}$	1.130	1.15	4.22
$\alpha_{10(2)}$	0.365	2.76	8.51
$\alpha_{20(1)}$	-0.684	3.11	11.90
$\alpha_{20(2)}$	-1.420	1.15	9.19
$\alpha_{30(1)}$	-0.909	3.37	11.30
$\alpha_{30(2)}$	-0.967	3.73	5.87
$\alpha_{40(1)}$	0.188	2.90	3.72
$\alpha_{40(2)}$	-0.039	3.46	6.59
$\alpha_{50(1)}$	1.310	4.23	14.60
$\alpha_{50(2)}$	1.320	4.21	9.52
$\alpha_{60(1)}$	-0.774	2.26	12.50
$\alpha_{60(2)}$	-1.140	2.21	4.46
α_{11}	0.006	1.32	3.95
α_{21}	-1.130	3.09	7.37
α_{31}	-1.090	4.75	13.80
α_{41}	-0.088	4.23	11.20
α_{51}	1.270	4.80	13.40
α_{61}	-1.560	2.54	4.26

model for the 2.5% and 50% quantiles estimated with precision +/- 0.005 at 95% confidence level after 10000 iterations with a 1000 burn-in.

The Geweke criterion showed that all parameters have converged. When the 2.5% quantile is estimated, dependence factor I is smaller than 5 for all parameters. That is not the case for the 50% quantile. Large values of the dependence factor I do not necessarily imply bad fit of the model. If nonconvergence is observed for some of the model parameters that means that the posterior distribution for that model parameters is not the correct one. Therefore the fit of the model will look bad when fitted values are checked when it might actually be good.

We also looked at the Heidelberger and Welch (1983) stationarity and interval width tests. All parameters of the model passed that test.

Table 6.2 gives the parameter estimates and standard errors for the one-

Table 6.2: Parameter Estimates and Standard Errors From the MCMC After 10000 Iterations and From the E-M Algorithm, for the One-factor Model, Environment Data.

	MCMC mean	E-M	Dif.	MCMC s.d.	E-M s.e.	Dif.
$\alpha_{10(1)}$	0.668	0.653	0.015	0.178	0.222	-0.044
$\alpha_{10(2)}$	3.594	3.537	0.057	0.342	0.341	0.001
$\alpha_{20(1)}$	2.528	2.443	0.085	0.392	0.342	0.050
$\alpha_{20(2)}$	6.025	5.809	0.216	0.770	0.872	-0.102
$\alpha_{30(1)}$	2.578	2.396	0.182	0.493	0.350	0.143
$\alpha_{30(2)}$	5.906	5.547	0.359	0.884	0.668	0.216
$\alpha_{40(1)}$	1.558	1.444	0.114	0.385	0.360	0.025
$\alpha_{40(2)}$	7.360	6.944	0.416	1.134	0.909	0.225
$\alpha_{50(1)}$	2.485	2.342	0.143	0.442	0.345	0.097
$\alpha_{50(2)}$	5.729	5.441	0.288	0.774	0.632	0.142
$\alpha_{60(1)}$	0.114	0.109	0.005	0.185	0.231	-0.046
$\alpha_{60(2)}$	2.543	2.506	0.037	0.282	0.315	-0.033
α_{11}	1.441	1.381	0.060	0.226	0.221	0.005
α_{21}	2.504	2.348	0.156	0.435	0.361	0.074
α_{31}	3.478	3.139	0.339	0.659	0.424	0.235
α_{41}	3.563	3.253	0.310	0.670	0.494	0.176
α_{51}	3.230	2.954	0.276	0.565	0.429	0.136
α_{61}	1.873	1.792	0.081	0.270	0.279	-0.009

Note. Dif. = Differences

factor model estimated using the E-M algorithm and from the Bayesian MCMC method. The results presented for MCMC have been obtained after 10000 iterations when convergence has been achieved according to most of the criteria and for most of the model parameters. Both estimation methods gave very similar results. MCMC gave slighter bigger values for all the model parameters, perhaps because it is aiming for the mean of a skew distribution, whereas maximum likelihood finds modes. The standard errors obtained from the standard deviation of the posterior distribution of each parameter tend to be greater than the asymptotic standard errors obtained from an approximation of the inverse of the information matrix.

6.3.2 Latent Trait Model For Binary Responses

This data set consists of 8445 individuals and eight binary items that measure women's independence in the society of Bangladesh. The data come
from the 1989 Fertility survey in Bangladesh (Huq & Cleland, 1990). The
manifest items analyzed are all binary and they ask married women whether
they could visit any part of the village, town, or city alone; go outside town
alone; talk to unknown men; go to the cinema alone; go shopping alone; go
to a club alone; participate in a political meeting; and visit a health center
or hospital alone.

We started the analysis by fitting a one-factor model which did not
give a satisfactory fit. The fit was checked using residuals from the bivariate and trivariate margins. The residuals are Pearson chi-square statistics
computed only for pairs and triplets of responses. Those values provide
information on how well the model predicts the two- and three-way margins. It must be noted that those residuals are not independent and the
chance of large values occurring by chance might increase with p. Each
residual can be treated individually as it follows a chi-square distribution
with one degree of freedom or a square of a standard normal. In that case
residuals with values greater than 4 will indicate a bad fit. The chi-square
statistic on the margins does not give a formal goodness-of-fit test of the
model but examination of the residuals reveals items or pair of items where
the model does not fit well. That might suggest collapsing categories or
omitting variables. The use of the one- and two-way margins to judge the
fit of the model has been discussed in Bartholomew and Tzamourani (1999)
for the latent trait model with binary items and in Jöreskog and Moustaki
(2002) for the latent trait model with ordinal responses.

Although the one-factor model is not a good fit we show in Table 6.3
the parameter estimates and standard errors for the MCMC after 2000
iterations and the E-M. Despite the bad fit of the model both methods
give very close results both for the parameter estimates and the standard
errors. We continued the analysis by fitting the two-factor latent trait
model. The two factor solution is not uniquely determined. An orthogonal
transformation of the factor loadings results in the same likelihood value.
To use BUGS we fixed the value of one of the loadings to make the factor
loadings unique. In our example, α_{31} was fixed to 1 to allow for a unique
solution.

The Geweke diagnostics after 2000 and 5000 iterations are presented in
Table 6.4. According to that criterion convergence has not been obtained
after 2000 iterations for some of the model parameters (see column 2 of
Table 6.4). The absolute value of the z-score is greater than 2 for thirteen
parameters. Iterations were increased to 5000 but convergence is still not

Table 6.3: **Parameter Estimates and Standard Errors From the MCMC Sampling Based on 2000 Iterations and From the E-M Algorithm, for the One-Factor Model, Women's Mobility Data.**

	MCMC mean	E-M	Dif.	MCMC s.d.	E-M s.e.	Dif.
α_{10}	2.275	2.282	-0.007	0.070	0.072	-0.002
α_{20}	-1.304	-1.301	-0.003	0.045	0.047	-0.002
α_{30}	1.547	1.546	0.001	0.044	0.044	0.000
α_{40}	-1.209	-1.205	-0.004	0.058	0.061	-0.003
α_{50}	-6.523	-6.591	0.068	0.278	0.285	-0.007
α_{60}	-4.476	-4.450	-0.026	0.158	0.144	0.014
α_{70}	-11.710	-10.715	-0.995	0.949	0.831	0.118
α_{80}	-4.815	-4.813	-0.002	0.169	0.137	0.032
α_{11}	2.098	2.105	-0.007	0.091	0.088	0.003
α_{21}	2.068	2.069	-0.001	0.069	0.071	-0.002
α_{31}	1.512	1.508	0.004	0.059	0.059	0.000
α_{41}	3.016	3.013	0.003	0.121	0.124	-0.003
α_{51}	4.044	4.094	-0.050	0.209	0.212	-0.003
α_{61}	3.232	3.211	0.021	0.139	0.132	0.007
α_{71}	7.033	6.393	0.640	0.622	0.544	0.078
α_{81}	3.069	3.068	0.001	0.139	0.111	0.028

Note. Dif. = Differences

achieved if we compare the first 10% of the chain with the last 50% (see column 3 of Table 6.4). However when we compare the first 20% of the chain with the last 50% (column 4) the absolute value of the z-score is less than 2 for all parameters but the ones that related to item 7 and the intercept of item 3. Column 5 of Table 6.4 gives the dependence factor I. According to that criterion most of the parameters are not accepted. The conclusion drawn is that different methods of checking the convergence of the chain can produce contradictory results. The same conclusion was drawn in example 1. Due to the contradictory results obtained from the different diagnostic tests one should increase the number of iterations even if that increases the computational estimation time significantly. In our case we stopped at 5000 iterations since the parameter estimates obtained are close to the ones produced by the E-M algorithm.

From Table 6.5 we see that the parameter estimates and the standard errors are very close. Item 7 has very large values on all parameters and big z-scores (see Table 6.4). That might be the reason why the Geweke

Table 6.4: Geweke Convergence Diagnostics (z-score) for the Two-Factor Model, Women's Mobility Data.

	Sample size 2000 z-score 10% vs 50%	Sample size 5000 z-score 10% vs 50%	Sample size 5000 z-score 20% vs 50%	Sample size 5000 factor I
α_{10}	0.208	1.370	0.942	6.51
α_{20}	-1.290	2.380	-1.530	7.36
α_{30}	-0.872	-1.930	-2.770	1.47
α_{40}	0.304	-2.590	-1.800	2.06
α_{50}	3.420	-1.870	-0.688	6.18
α_{60}	-6.860	-2.270	-1.100	20.40
α_{70}	40.000	22.200	20.300	30.80
α_{80}	-3.100	-3.500	-1.230	11.00
α_{11}	0.424	1.520	1.320	5.79
α_{21}	1.470	-3.700	1.060	9.64
α_{31}	1.470	-0.643	-0.908	4.29
α_{41}	1.430	0.518	0.368	6.51
α_{51}	1.700	-1.330	-0.438	6.77
α_{61}	3.800	-1.580	0.301	5.52
α_{71}	-7.120	-11.600	-10.300	22.60
α_{81}	4.860	-1.520	-0.109	5.83
α_{22}	-2.800	-0.058	-0.131	5.38
α_{32}	-3.080	-0.118	-0.256	4.88
α_{42}	-3.160	0.951	0.844	7.01
α_{52}	-4.270	2.100	0.950	9.09
α_{62}	6.560	2.540	1.160	10.40
α_{72}	-36.100	-22.100	-20.700	16.10
α_{82}	0.976	3.360	1.130	5.73

criterion gave a large z-score.

Table 6.5: Parameter Estimates and Standard Errors From MCMC After 5000 Iterations and From the E-M Algorithm, for the Two-Factor Model, Women's Mobility Data.

	MCMC mean	E-M	Dif.	MCMC s.d.	E-M s.e.	Dif.
α_{10}	2.644	2.663	-0.019	0.142	0.179	-0.037
α_{20}	-1.612	-1.582	-0.03	0.090	0.087	0.003
α_{30}	1.551	1.555	-0.004	0.043	0.045	-0.002
α_{40}	-1.196	-1.171	-0.025	0.062	0.061	0.001
α_{50}	-6.541	-6.579	0.038	0.273	0.300	-0.027
α_{60}	-5.331	-5.109	-0.222	0.283	0.269	0.014
α_{70}	-19.27	-17.245	-2.025	2.901	94.815	-91.914
α_{80}	-4.977	-4.936	-0.041	0.179	0.166	0.013
α_{11}	2.429	2.462	-0.033	0.206	0.282	-0.076
α_{21}	2.495	2.475	0.02	0.209	0.210	-0.001
α_{31}	1.238	1.246	-0.008	0.076	0.084	-0.008
α_{41}	1.969	1.975	-0.006	0.121	0.160	-0.039
α_{51}	1.967	1.985	-0.018	0.205	0.229	-0.024
α_{61}	1.260	1.323	-0.063	0.169	0.234	-0.065
α_{71}	3.504	2.203	1.301	0.688	0.430	0.258
α_{81}	1.486	1.511	-0.025	0.149	0.172	-0.023
α_{21}	1.000	0.976	0.024	0.000	0.169	-0.169
α_{22}	1.374	1.324	0.05	0.108	0.151	-0.043
α_{32}	0.868	0.857	0.011	0.082	0.100	-0.018
α_{42}	2.294	2.264	0.03	0.134	0.167	-0.033
α_{52}	3.523	3.570	-0.047	0.202	0.216	-0.014
α_{62}	3.802	3.604	0.198	0.260	0.237	0.023
α_{72}	11.26	10.011	1.249	1.778	58.023	-56.245
α_{82}	2.828	2.798	0.03	0.164	0.172	-0.008

Note. Dif. = Differences

6.3.3 Latent Trait Model for Binary Responses With Missing Values

The data set used here is a random subset of the 1989 Bangladesh Fertility Survey. Here, we take a subsample of 540 married women and we use only the first five indicators (we have excluded the indicators *club alone*, *political meeting*, and *health center*). The reason for excluding those three items is so that the five remaining attitudinal items are unidimensional before creating the pseudo response indicators and introducing the response propensity latent dimension.

For this example we alter some responses to check how our missing value model works. A proportion of 'no' responses to each attitudinal item, but none of the 'yes' responses was changed to missing. By doing that we imply that women who do not respond are more likely to respond 'no' to the five attitude items and, as a result, we have artificially forced the nonresponse to depend on attitude. The model with missing values was fitted to those five items. This data set was analyzed in Moustaki and Knott (2002b).

In Table 6.6 we show the Geweke diagnostic criterion for the model after 4000, 9000 and 14000 iterations and the dependence factor I after 9000 and 14000 iterations. The Geweke criterion given in the table compares the first 10% of the chain with the last 50%. Even after 14000 iterations, the z-scores and the dependence factor I show evidence of non convergence for most of the parameters. Estimation problems are encountered with BUGS when we tried to increase the number of iterations and therefore we report the results obtained after the 14000 iterations. That might well be the reason for the differences observed between the E-M and the MCMC parameter estimates (see Table 6.7). Bigger differences are found on the loadings of the second factor (response propensity).

6.3.4   Computational Aspects

Table 6.8 gives the times needed by BUGS and GENLAT to obtain the parameter estimates and standard errors for the different examples. It is clear that the E-M estimation method is much faster than the MCMC method. The MCMC slows down significantly with the increase in the number of factors from one to two.

6.4   Conclusion

The parameter estimates for the two examples without missing values are close both from the MCMC approach and the E-M. The convergence diagnostics for the missing values example suggested more iterations but BUGS failed to estimate the model when the number of iterations increased.

Overall the E-M is faster than the MCMC algorithm. The E-M provides asymptotic standard errors where the MCMC approach computes the whole posterior distribution of each parameter and the standard errors are the standard deviations of those posterior distributions. We used BUGS for

Table 6.6: Geweke Convergence Diagnostic (z-score) and the Dependence Factor I for the Missing Values Model.

	z-score			I	
	4000	9000	14000	9000	14000
α_{10}	0.412	1.980	0.732	1.22	1.16
α_{20}	-1.530	1.370	-0.144	2.88	4.03
α_{30}	2.580	0.526	-1.410	1.20	1.19
α_{40}	7.000	-1.570	-0.163	22.80	21.20
α_{50}	14.400	6.820	1.360	44.60	53.50
α_{60}	-3.770	2.280	-4.530	3.93	5.60
α_{70}	-4.030	-3.580	0.499	1.58	1.47
α_{80}	-1.130	-4.840	-3.650	3.40	3.32
α_{90}	1.900	-1.290	3.650	5.28	5.88
α_{10}	1.320	-0.118	-1.760	2.82	4.26
α_{11}	0.014	0.822	1.620	2.88	2.59
α_{21}	4.060	-0.551	0.926	1.25	2.10
α_{31}	1.540	-1.430	-1.900	1.44	2.30
α_{41}	-6.220	2.590	0.478	10.60	11.80
α_{51}	-14.600	-6.520	-1.150	23.30	22.10
α_{61}	-3.370	1.750	-4.110	3.93	3.78
α_{71}	-2.880	-2.610	-0.920	1.40	1.33
α_{81}	-2.000	-3.910	-2.390	3.20	3.12
α_{91}	1.780	-1.070	2.660	4.15	4.83
$\alpha_{10,1}$	-1.450	-0.755	-1.520	2.68	2.79
α_{62}	-3.200	2.550	-4.650	12.00	8.54
α_{72}	-3.710	-3.520	0.905	3.62	2.61
α_{82}	0.803	-4.090	-3.400	1.39	2.09
α_{92}	1.330	-1.260	3.780	1.10	2.28
$\alpha_{10,2}$	3.570	0.440	-1.360	2.59	2.43

Table 6.7: Parameter Estimates and Standard Errors From the MCMC Sampling After 14000 Iterations and From the E-M Algorithm for the Missing Values Model.

	MCMC mean	E-M	Dif.	MCMC s.d.	E-M s.e.	Dif.
α_{10}	1.705	1.683	0.022	0.181	0.190	-0.009
α_{20}	-1.195	-1.252	0.057	0.162	0.190	-0.028
α_{30}	1.422	1.400	0.022	0.159	0.162	-0.003
α_{40}	-2.925	-3.443	0.518	1.491	43.833	-42.342
α_{50}	-10.130	-12.278	2.148	3.966	261.589	-257.623
α_{60}	8.889	10.012	-1.123	5.421	40.425	-35.004
α_{70}	5.629	4.373	1.256	2.500	0.703	1.797
α_{80}	6.958	5.793	1.165	2.085	0.993	1.092
α_{90}	8.225	5.904	2.321	3.745	1.499	2.246
α_{10}	4.779	4.234	0.545	1.259	0.538	0.721
α_{11}	1.592	1.587	0.005	0.246	0.260	-0.014
α_{21}	1.813	1.893	-0.08	0.226	0.284	-0.058
α_{31}	1.456	1.437	0.019	0.209	0.219	-0.010
α_{41}	10.570	10.029	0.541	4.993	113.782	-108.789
α_{51}	8.387	10.155	-1.768	3.445	224.843	-221.398
α_{61}	2.552	2.857	-0.305	1.544	10.989	-9.445
α_{71}	1.165	0.987	0.178	0.599	0.407	0.192
α_{81}	2.167	1.880	0.287	0.816	0.545	0.271
α_{91}	3.460	2.343	1.117	2.047	0.988	1.059
$\alpha_{10,1}$	1.703	1.536	0.167	0.508	0.377	0.131
α_{12}	0.000					
α_{22}	0.000					
α_{32}	0.000					
α_{42}	0.000					
α_{52}	0.000					
α_{62}	3.957	4.795	-0.838	3.230	22.44	-19.21
α_{72}	1.651	0.842	0.809	1.562	0.710	0.852
α_{82}	1.521	0.896	0.625	1.168	0.850	0.318
α_{92}	1.009	0.267	0.742	0.989	1.133	-0.144
$\alpha_{10,2}$	1.046	0.664	0.382	0.873	0.649	0.224

Note. Dif. = Differences

Table 6.8: Time Needed for MCMC and E-M to Converge for the Three Examples.

Conditions				Estimation Method			
n	p	c	q	MCMC		E-M	
				CPU sec.	Iterations	CPU sec.	Iterations
291	6	3	1	750	10000	8	366
8445	8		1	4787	2000	1	54
8445	8		2	24480	5000	4	258
540	10		2	3060	14000	4	46

other types of data such as nominal, metric and mixed data and it worked satisfactorily. As the complexity of the model increases because of large numbers of factors and items BUGS has problems converging. The same has not been found for the E-M estimation method.

The models compared in this chapter are large in terms of the number of parameters that need to be estimated. The posterior distribution of the whole parameter space is too complex to be calculated.

One of the big issues in MCMC is the convergence of the parameter estimates. We used some of the diagnostics available in CODA to check the convergence of our models. Contradictory results often arise. We also noticed in the second example that, although some of the diagnostics suggest nonconvergence, the parameter estimates are close to the ones obtained with the E-M.

In all the models we have used noninformative priors for the model parameters (thresholds, difficulty parameters, and factor loadings). For the case where the parameter estimates behave strangely (become very large) one could try to use informative priors.

References

Agresti, A. (1990). *Categorical data analysis.* New York: Wiley.

Albert, J. H. (1992). Bayesian estimation of normal ogive item response curves using Gibbs sampling. *Journal of Educational Statistics, 17*, 251-269.

Albert, J. H., & Chib, S. (1993). Bayesian analysis of binary and polychotomous response data. *Journal of the American Statistical Association, 88*, 669-679.

Bartholomew, D. J., & Knott, M. (1999). *Latent variable models and factor analysis* (2nd ed.). London: Arnold.

Bartholomew, D. J., & Tzamourani, P. (1999). The goodness-of-fit of latent trait models in attitude measurement. *Sociological Methods and Research, 27*, 525-546.

Best, N. G., Cowles, M. K., & Vines, S. K. (1996). *Coda manual version 0.30.* Retrieved from *http://www.biostat.ucsf.edu/docs/bugs/cdaman03.pdf*

Bock, R. D., & Aitkin, M. (1981). Marginal maximum likelihood estimation of item parameters: application of an EM algorithm. *Psychometrika, 46*, 443-459.

Bock, R. D., Gibbons, R., & Muraki, E. (1988). Full-information item factor analysis. *Applied Psychological Measurement, 12*, 261-279.

Dunson, D. B. (2000). Bayesian latent variable models for clustered mixed outcomes. *Journal of the Royal Statistical Society, Series B, 62*, 355-366.

Geweke, J. (1992). Evaluating the accuracy of sampling-based approaches to calculating posterior moments. In J. M. Bernardo, J. O. Berger, A. P. Dawid & A. F. M. Smith (Eds.), *Bayesian statistics 4* (pp. 169-188). Oxford, England: Clarendon Press.

Heidelberger, P., & Welch, P. (1983). Simulation run length control in the presence of an initial transient. *Operations Research, 31*, 1109-1144.

Huq, N. M., & Cleland, J. (1990). *Bangladesh fertility survey, 1989.* Dhaka, Bangladesh: National Institute of Population Research and Training (NIPORT).

Jöreskog, K. G., & Moustaki, I. (2002). Factor analysis of ordinal variables: a comparison of three approaches. *Multivariate Behavioral Research, 36*, 347-387.

Knott, M., & Albanese, M. T. (1993). POLYMISS: A computer program for fitting a one- or two- factor logit-probit latent variable model to polytomous data when observations may be missing. Unpublished manuscript, London School of Economics and Political Science, England & Universidade Federal do Rio Grande do Sul, Brazil.

Knott, M., Albanese, M. T., & Galbraith, J. (1990). Scoring attitudes to abortion. *The Statistician, 40*, 217-223.

Louis, T. A. (1982). Finding the observed information matrix when using the EM algorithm. *Journal of the Royal Statistical Society, Series B, 44*, 226-233.

Mislevy, R. J., & Bock, R. D. (1990). BILOG3: Item analysis and test scoring with binary logistic models. [Computer software.] Chicago: Scientific Software Inc.

Moustaki, I. (1996). A latent trait and a latent class model for mixed observed variables. *British Journal of Mathematical and Statistical Psychology, 49*, 313-334.

Moustaki, I. (2000a). A latent variable model for ordinal variables. *Applied Psychological Measurement, 24*, 211-223.

Moustaki, I. (2000b). GENLAT: A computer program for fitting a one- or two- factor latent variable model to categorical, metric and mixed observed items with missing values. [Computer software.] Statistics Department, London School of Economics and Political Science. Retrieved, October 5, 2003, from
http://stats.lse.ac.uk/knott/software/latbug/.

Moustaki, I. (2003). A general class of latent variable models for ordinal manifest variables with covariate effects on the manifest and latent variables. *British Journal of Mathematical and Statistical Psychology, 56*, 337-358.

Moustaki, I., & Knott, M. (2000a). Generalized latent trait models. *Psychometrika, 65*, 391-411.

Moustaki, I., & Knott, M. (2000b). Weighting for item non-response in attitude scales by using latent variable models with covariates. *Journal of the Royal Statistical Society, Series A, 163*, 445-459.

Moustaki, I., & O'Muircheartaigh, C. (2002). Locating 'don't know,' 'no answer,' and middle alternatives on an attitude scale: A latent variable approach. In G. Marcoulides & I. Moustaki (Eds.), *Latent variable and latent structure models* (pp. 15-41). Mahwah, NJ: Erlbaum.

Muraki, E., & Carlson, E. (1995). Full-information factor analysis for polytomous item responses. *Applied Psychological Measurement, 19*, 73-90.

O'Muircheartaigh, C., & Moustaki, I. (1999). Symmetric pattern models: A latent variable approach to item non-response in attitude scales. *Journal of the Royal Statistical Society, Series A, 162*, 177-194.

Patz, R. J., & Junker, B. W. (1999a). Applications and extensions of MCMC in IRT: Multiple item types, missing data, and rated responses. *Journal of Educational and Behavioral Statistics, 24*, 342-366.

Patz, R. J., & Junker, B. W. (1999b). A straightforward approach to Markov chain Monte Carlo methods for item response models. *Journal of Educational and Behavioral Statistics, 24*, 146-178.

Rabe-Hesketh, S., Pickles, A., & Skrondal, A. (2001). GLLAMM: Stata program to fit generalised linear latent and mixed models (Manual Technical Report 2001). [Computer software.] London: King's College London, Institute of Psychiatry.

Raftery, A. L., & Lewis, S. M. (1992a). Comment: One long run with diagnostics: Implementation strategies for Markov chain Monte Carlo. *Statistical Science, 7*, 493-497.

Raftery, A. L., & Lewis, S. M. (1992b). How many iterations in the Gibbs sampler? In J. M. Bernardo, J. O. Berger, A. P. Dawid & A. F. M. Smith (Eds.), *Bayesian statistics 4* (pp. 763-773). Oxford, England: Clarendon Press.

Samejima, F. (1969). Estimation of latent ability using a response pattern of graded scores. *Psychometrika, Monograph Supplement, 17*.

Sammel, R. D., Ryan, L. M., & Legler, J. M. (1997). Latent variable models for mixed discrete and continuous outcomes. *Journal of the Royal Statistical Society, Series B, 59*, 667-678.

Spiegelhalter, D., Thomas, A., Best, N. G., & Gilks, W. (1996). BUGS 0.5: Bayesian inference using Gibbs sampling. [Computer software.] Cambridge, England: Institute of Public Health, MRC Biostatistics Unit.

Thissen, D. (1991). MULTILOG: Multiple, categorical items analysis and test scoring using item response theory. [Computer software.] Chicago: Scientific Software, Inc.

Wilson, D., Wood, R. L., & Gibbons, R. (1991). TESTFACT 2. [Computer software.] Chicago: Scientific Software International.

Chapter 7

Logistic Models
for Single-Subject Time Series

Peter W. van Rijn and Peter C.M. Molenaar
University of Amsterdam

7.1 Introduction

Statistical methodology developed in psychology is mostly applied to a collection of individuals rather than to a single person (Kratochwill, 1978, p. 3). The development of psychological testing methods in the first half of the 20th-century put the individual on the background as the initial objective was to differentiate among individuals. Keeping this in mind, the deliberate focus on advancement of analysis techniques based on variation between individuals (inter-individual variation, IEV) instead of variation within a single individual seems tenable. However, models for time-dependent variation of a single individual (intra-individual variation, IAV) have been widely available for some time. The discovery of the intrinsically stochastic time-dependent behavior within grains of pollen (Brownian motion) led to the development of appropriate models for single systems in the beginning of the 20th century. In this regard, the absence of a pure $N = 1$ perspective in psychometrics might be perceived as startling.

It is not to say that examples of analyses of IAV are missing in the psychometric literature. The measurement of (individual) change, for example, is a branch of psychometrics with a relatively long history. An early overview of problems encountered in measuring change can be found in Harris (1962). In that book, a single-subject analysis of multivariate time series is described by Holtzmann, who stresses that psychologists should study this type of analysis because of the increasing importance of time series in other branches of science such as econometrics and biometrics (Holtzmann, 1962, p. 199). More recently, Nesselroade and Schmidt Mc-Collam (2000) advocate analyzing IAV in the context of developmental processes in psychology and Collins and Sayer (2001) provide an overview of newly developed methods for the analysis of change.

Apart from the historical development, it is difficult to find an explicit and convincing rationale for the one-sided focus on IEV in contemporary psychometrics. The restriction to IEV appears to be considered to be an almost self-evident consequence of the scientific ideal to strive for general nomothetic knowledge. The science of psychology should involve theories and laws that apply to all human subjects. Such nomothetic knowledge would seem to be ill suited by intensive study of single subjects, because results thus obtained may not be generalizable in the intended sense. Despite its possible appeal, we argue that this kind of rationale is incorrect in many instances by making use of a set of well-known mathematical theorems.

In our criticism of the one-sided focus on IEV we do not take issue with the ideal of nomothetic knowledge, that is, the search for psychological theories and laws that apply to all human subjects. Our criticism only concerns the assumption that theories and laws based on analysis of IEV apply to each human subject, and thus, would hold for IAV. To obtain valid theories and laws about IAV, one cannot generalize results derived from IEV, but one has to study IAV in its own right. That is the implication of the mathematical theorems we refer to. Having available the results of a sufficient number of individual analyses of IAV, one then can search for general characteristics by means of standard inductive techniques. If successful this will yield valid nomothetic knowledge about the structure of IAV, that is, nomothetic theories and laws about idiographic (individual) processes.

This chapter is divided into two parts. The first part starts out with a description of analyses for IEV and IAV and the condition under which there exists a relationship between the two types of analysis, namely ergodicity. This condition is explained in the context of psychometrics.

Having thus set the stage for serious consideration of IAV, the second part of this chapter discusses latent variable models for single-subject time series data. Special attention is given to the logistic model for multivariate dichotomous time series which can be seen in its simplest form as a dynamic variant of the Rasch model. It must be stated that logistic models for repeated measurements have been discussed by various authors (e.g., Kempf, 1977; Fischer 1983, 1989; Verhelst & Glas, 1993; Agresti, 1997). In most of these applications, however, the IAV nature of single-subject data is not emphasized as much as in the present chapter. To the best of our knowledge, an application of the Rasch model to IAV is presented here for the first time in the literature.

7.2 Ergodicity: the Relation Between IEV and IAV

In its basic form, standard statistical analysis in psychology proceeds by drawing a sample of subjects, assessing their scores on selected measuring instruments, and then computing statistics by taking appropriate averages over the scores of all available subjects. If all subjects would yield the same score, statistical analysis would be severely reduced. Hence, it is the manner in which scores vary across subjects, IEV, which provides the information for the analysis. In contrast, in time series analysis the same individual subject is repeatedly measured, and statistics are computed by taking appropriate averages over her scores obtained at all measurement occasions. Hence it is the manner in which a subject's scores vary across measurement occasions, IAV, which provides the information for time series analysis.

We already indicated that psychometricians are mainly interested in analyses of IEV. A vivid illustration of this tendency can be found in the classic treatise of test theory by Lord and Novick (1968). They define the concept of true score of a person as the mean of the distribution of scores obtained by independent repeated measurement of this person. This is obviously a definition in terms of IAV. Lord and Novick then remark that repeated measurement of the same person will affect this person's state and give rise to fatigue, habituation, or other confounding effects. They conclude that therefore, instead of measuring one person a large number of times, test theory has to be based on the alternative paradigm in which a large number of persons is measured once or twice. The shift to the latter alternative paradigm implies that test theory is based on analysis of IEV.

Notwithstanding that confounding factors such as habituation and fatigue might complicate the implementation of a purely IAV based test theory, a reference to such contingent states of affairs cannot be given as a reason for the impossibility of this whole paradigm. In addition, Lord and Novick (1968, p. 32) state that the definition of true score in terms of IAV would be better suited for individual assessment than an IEV based test theory that is meant to differentiate among individuals. It might therefore

be expected that psychological tests constructed on the basis of analysis of
IEV perform suboptimal when applied for the purpose of individual prediction. However, a task that still awaits further elaboration is the assessment
of situations in which such test performance is suboptimal.

The urgency to determine the performance of standard tests in the context of individual assessment and prediction becomes even more pressing
because strong reasons can be given for the conjecture that the differences
between analysis of IEV and IAV go deeper than a mere difference in degree
of success in the context of individual prediction. The reasons we have in
mind are of two kinds: Implications of ergodic theorems and results from
mathematical biology suggesting the presence of substantial heterogeneity in human populations. Ergodic theory concerns the characterization
of stochastic processes for which analysis of IEV and IAV yield the same
results (e.g., Petersen, 1983). So for ergodic processes the one-sided focus on analysis of IEV does not present any problem, because results thus
obtained also are valid for individual assessment and prediction of IAV. Unfortunately, however, the criteria for ergodicity are very strict and involve
the absence of any time-dependent changes in the distributional characteristics of a stochastic process. Hence all developmental, learning and adaptive
processes do not obey the criteria for ergodicity and for these classes of nonergodic processes there may not exist any lawful relationship between IEV
and IAV.

Related to ergodicity is the notion of stationarity, which concerns the
distributional characteristics of a single realization of a stochastic process.
Stationarity amounts to the absence of time-dependent changes in distributional characteristics and is a necessary condition for ergodicity.[1] A simple
example of a stationary process is the coin-tossing model in which the outcome of each toss is the variable of interest, heads and tails are scored $+1$
and -1, and tosses are obtained independently. It is easily seen that the
distributional properties are time-invariant. Now if we change the variable
of interest to the sum of the outcomes, we no longer have a stationary
process as the variance increases with time.

Even if the distributional characteristics of a stochastic process are invariant in time, that is, the process is stationary, it still may be nonergodic.
The key difference between stationarity and ergodicity concerns the unique-
ness of the so-called equilibrium distribution of a stochastic process, that
is, the distribution of the values of a stochastic process as time increases
without bound. Each stationary process gives rise to an equilibrium distribution, but this equilibrium distribution may not be unique. Only if the
process is ergodic, then this is necessary and sufficient for its equilibrium

---

[1] Note that strict stationarity is mentioned here (see, e.g., Hamilton, 1994, pp. 45-46).

distribution to be unique (cf. Mackey, 1992, Theorem 4.6). Hence stationary processes are nonergodic if they display a moderate kind of heterogeneity: Their equilibrium distribution is not unique. Notice that this is the kind of nonergodicity known from Markov chain theory (e.g., Kemeny, Snell, & Knapp, 1966). Already the presence of this moderate form of heterogeneity with respect to the equilibrium distribution implies the possibility of a lack of lawful relationships between IEV and IAV.

An example of a stationary and ergodic process can be obtained when we apply the first coin-tossing model to each coin in a bag of fair coins. The process is ergodic as the values of the tosses of each coin are governed by the same unique equilibrium distribution. However, if the bag contains unfair coins and the average probability of heads remains $\frac{1}{2}$, then we have a nonergodic, stationary process. This can be inferred from the fact that there exist equilibrium distributions for different coins, that is, sequences obtained from single coins are stationary. Yet, there exists no unique equilibrium distribution which governs sequences obtained from randomly drawn coins. Hence, the process is nonergodic.

There are strong indications that heterogeneity in human populations may be much more pervasive, transcending the moderate forms associated with nonergodicity. Mathematical theory about biological pattern formation (cf. Murray, 1993) and nonlinear epigenetics (Edelman, 1987) shows that growth processes are severely underdetermined by genetic and environmental influences. Consequently, growth processes have to be self-organizing to accomplish their tasks. In particular the maturation of the central nervous system results from self-organizing epigenetic processes. Self-organization, however, gives rise to substantial endogenous variation that is independent from genetic and environmental influences (Molenaar, Boomsma, & Dolan, 1993; Molenaar & Raijmakers, 1999). For instance, homologous neural structures on the left-hand and right-hand side of the same individual (IAV) can differ as much as the left-hand side of this neural structure in different individuals (IEV). Insofar as the activity of such heterogeneous neural structures is associated with the performance on psychological tests, this performance can be expected to be heterogeneous in much stronger forms than is the case with nonergodicity.

It has been shown by means of simulation experiments as well as mathematical proof (Molenaar, Huizenga, & Nesselroade, 2003; Kelderman & Molenaar, in press) that standard factor analysis of IEV is insensitive to the presence of substantial heterogeneity. For instance, it is an assumption of the standard factor model that factor loadings are invariant (fixed) across subjects. If, however, these factor loadings would in reality vary randomly across subjects (a grave violation of the assumption of fixed factor loadings), then the standard factor model still fits satisfactorily. There appears to be only one principled way in which the presence of such heterogeneity can be detected, namely by carrying out replicated factor analyses of IAV (dynamic factor analysis of multivariate time series; cf. Molenaar, 1985) and then compare the solutions thus obtained for distinct subjects.

In closing this section, it is reiterated that in general one cannot expect lawful relationships to exist between the structure of IEV and the structure of IAV. Such relationships only obtain under the restrictive condition that the processes concerned are ergodic. For nonergodic processes and in cases where human subjects are heterogeneous in even more pervasive ways (e.g., each subject having its personal factor model with its own distinct number of factors, factor loading pattern and/or specific variances), the use of IAV paradigms is mandatory. To accomplish this, appropriate time series analysis extensions of standard statistical techniques are required. Brillinger (1975) presents a rigorous derivation of time series analogues of all standard multivariate techniques (analysis of variance, regression analysis, principal component analysis, canonical correlation analysis). In the next section, we present an overview of time series analogues of latent variable models.

7.3   Latent Variable Models for IAV

From a general point of view, a stochastic process can be interpreted as a random function. That is, as an ensemble of time-dependent functions on which a probability measure is defined (cf. Brillinger, 1975, section 2.11). Each time-dependent function of this ensemble is called a *trajectory* (or *realization*). Even if information is available about the entire past of a stochastic process up to some time t, then exact prediction for the next time point still is impossible. Each trajectory in an ensemble extends over the entire time axis. An observed time series, that is, the particular stretch of values obtained by repeated measurement of a single subject, constitutes a randomly drawn trajectory from the ensemble, where this trajectory is clipped by a time window with width equal to the period of repeated measurement. In what follows we denote a stochastic process by y_t and an observed time series thereof by y_t, $t = 1, \ldots, T$. We acknowledge that this notation is not entirely correct, but it is convenient and customary.

A subset of latent variable models for IAV is obtained by replacing all random variables in a standard latent variable model by stochastic processes. Bartholomew (1987) gave a useful classification of standard latent variable models based on two features: Whether the observed variable is continuous or discrete and whether the common latent variable is continuous or discrete. This classification will be followed in our overview of latent variable models for IAV. There is an additional third feature which has to be considered for latent variable models for IAV, namely whether the time dimension is continuous or discrete. We, however, restrict attention to models in discrete time only, as this is sufficient for our present purposes.

If both the observed variable and the common latent variable are continuous, the latent variable model is classified as a factor model. Replacement of all random variables in the factor model by continuous stochastic processes yields the state-space model: $y_t = Z_t \alpha_t + \epsilon_t$, wherein y_t is the observed continuous n-variate process, α_t is a common m-variate latent process (also called *state process*), and ϵ_t is a n-variate measurement error process. Statistical analysis of the state-space model is well developed and is treated in several text books (e.g., Durbin & Koopman, 2001). Hamaker, Dolan, and Molenaar (2003) discuss applications of the state-space model in psychological research.[2]

If both the observed variable and the common latent variable are discrete, the latent variable model is classified as a latent class model. Replacement of all random variables in the latent class model by discrete stochastic processes yields the hidden Markov model (cf. Elliott, Aggoun, & Moore, 1995). Visser, Raijmakers, and Molenaar (2000) present applications of hidden Markov modelling in psychological research.[3] If the observed variable is continuous and the common latent variable is discrete, the latent variable model is classified as a latent profile model (Bartholomew, 1987; Molenaar & Von Eye, 1994). Replacement of the observed variable by a continuous stochastic process and the common latent variable by a discrete stochastic process yields a variant of the hidden Markov model (Elliott et al., 1995).

If the observed variable is discrete and the common latent variable is continuous, the latent variable model is classified as a generalized linear model. Replacement of all random variables in the generalized linear model by discrete (observed) and continuous (latent) stochastic processes yields the dynamic generalized linear model (Fahrmeir & Tutz, 2001).

We focus on a subset of dynamic generalized linear models. That is, models in which the observed process is dichotomous, related to the continuous latent process through the logistic response function.

---

[2]Software for the fit of state-space models can be downloaded from *http://users.fmg.uva.nl/cdolan/*.

[3]Appropriate software can be found at *http://users.fmg.uva.nl/ivisser/hmm*.

7.4   A Logistic Model for Dichotomous Time Series

Dichotomous (or binary) time series can be modelled in various ways. Regression models for dichotomous time series are discussed in detail in Kedem and Fokianos (2002) and in Fahrmeir and Tutz (2001). Our focus is on modelling dichotomous time series using latent variables which is comparable to the modelling of dichotomous variables in item response theory (see, e.g., Hambleton & Swaminathan, 1985). Because the latent variable is replaced by a stochastic process, it can be seen as a dynamic extension of item response modelling. As stated before, this approach is not entirely new although the emphasis on the modelling of IAV in this sense is novel. Modelling is pursued following Fahrmeir and Tutz (2001), that is, by specifying the distributional model, the response function, the linear predictor, and the transitional model.

7.4.1   General Outline

Consider the situation in which we have a dichotomously scored, multivariate time series, that is, an n-dimensional observation vector $\mathbf{y}_t$ such that $\mathbf{y}_t \in \{0,1\}^n$, at each time point $t = 1, \ldots, T$. Each single univariate observation y_{it}, $i = 1, \ldots, n$, follows a Bernoulli distribution with parameter π_{it} as the probability of obtaining a score one, given by

$$y_{it} \sim B(\pi_{it}) = \pi_{it}^{y_{it}}(1 - \pi_{it})^{1-y_{it}}. \tag{7.1}$$

This probability is modelled by inserting the linear predictor η_{it} into the logistic response function, resulting in

$$\pi_{it} = \frac{\exp(\eta_{it})}{1 + \exp(\eta_{it})}. \tag{7.2}$$

Next, the n-dimensional linear prediction vector $\boldsymbol{\eta}_t$ is constructed by linking the $n \times m$ design matrix $\mathbf{Z}_t$ with the m-dimensional latent state vector $\boldsymbol{\alpha}_t$, $t = 0, 1, \ldots, T$

$$\boldsymbol{\eta}_t = \mathbf{Z}_t \boldsymbol{\alpha}_t.$$

The linear transition equation, which relates states at $t - 1$ to t through the $m \times m$ transition matrix $\mathbf{F}_t$, is given by

$$\boldsymbol{\alpha}_t = \mathbf{F}_t \boldsymbol{\alpha}_{t-1} + \mathbf{R}_t \boldsymbol{\xi}_t, \quad t = 1, \ldots, T.$$

The state vector $\boldsymbol{\alpha}_t$ is allowed to contain time-invariant elements. The $m \times p$ selection matrix $\mathbf{R}_t$ is assumed to be a subset of the columns of the

m-dimensional identity matrix $\mathbf{I}_m$, so that it associates the elements of the p-dimensional disturbance vector $\boldsymbol{\xi}_t$ with the p time-varying elements of the state vector (Durbin & Koopman, 2001, p. 38).[4] The elements of $\boldsymbol{\xi}_t$ are often called innovations. The initial state $\boldsymbol{\alpha}_0$ and the disturbance vector $\boldsymbol{\xi}_t$ are normally distributed as

$$\boldsymbol{\alpha}_0 \sim \mathrm{N}(\mathbf{a}_0, \mathbf{R}_0 \mathbf{Q}_0 \mathbf{R}_0'), \quad \text{and} \quad \boldsymbol{\xi}_t \sim \mathrm{N}(\mathbf{0}, \mathbf{Q}_t).$$

The covariance of $\boldsymbol{\alpha}_0$ is represented in this manner so that $\mathbf{Q}_0$ is nonsingular, which is somewhat more advantageous (Durbin & Koopman, 2001, p. 38). Note that $\mathbf{a}_0$, $\mathbf{Q}_0$, and $\mathbf{Q}_t$ are hyperparameters to be estimated (see Appendix).

The following three assumptions are stated to completely specify the model in terms of densities. The first assumption is that current observations are dependent on current states

$$p(\mathbf{y}_t | \boldsymbol{\alpha}_t, \boldsymbol{\alpha}_{t-1}, \dots,, \boldsymbol{\alpha}_0, \mathbf{y}_{t-1}, \mathbf{y}_{t-2}, \dots, \mathbf{y}_1) = p(\mathbf{y}_t | \boldsymbol{\alpha}_t, \mathbf{y}_{t-1}, \mathbf{y}_{t-2}, \dots, \mathbf{y}_1).$$

The second assumption is that the state process is Markovian

$$p(\boldsymbol{\alpha}_t | \boldsymbol{\alpha}_{t-1}, \dots, \boldsymbol{\alpha}_0) = p(\boldsymbol{\alpha}_t | \boldsymbol{\alpha}_{t-1}).$$

Finally, and in addition to assumption one, it is assumed that the multivariate observations are independent given the current state

$$p(\mathbf{y}_t | \boldsymbol{\alpha}_t) = \prod_{i=1}^{n} p(y_{it} | \boldsymbol{\alpha}_t).$$

Because the specific contents of the state and disturbance vector can be freely chosen, a variety of latent processes can be captured with the current representation. Depending on the hypothesized dynamic constellation of the latent process, one can choose between for instance, autoregressive processes, moving average processes, and random walks (for a description, see, e.g., Hamilton, 1994). In addition, trends and cyclic change parameters can be included in the current representation. For now, we consider the latent process to be a random walk, so that the transition matrix F_t is fixed. For other types of processes, the transition matrix can contain hyperparameters, for example, autoregression parameters.

The model can be extended to more than one person ($N > 1$), more than one latent process ($p > 1$), and also to polytomous variables. For now, however, the interest lies in $N = 1$ and as results of analyses of dichotomous time series with this type of models are scarce, we next consider a simple, yet illustrative modelling example.

---

[4]It is stressed that $m = p + n$ does not necessarily follow.

7.4.2   A Dynamic Rasch Model

We now illustrate how a dynamic variant of the Rasch model can be obtained. We choose to let the state vector $\boldsymbol{\alpha}_t$ consist of two parts. The first part describes a person's latent process denoted by the p-dimensional vector $\boldsymbol{\theta}_t$. The second part consists of n threshold parameters, denoted by $\boldsymbol{\beta}_t$. So, we have $\boldsymbol{\alpha}_t = (\boldsymbol{\theta}'_t, \boldsymbol{\beta}'_t)'$ and $m = p + n$. For simplicity and sufficiency for present purposes, the following model parameters and matrices are considered time invariant: the design matrix ($\mathbf{Z}$), the threshold parameters ($\boldsymbol{\beta}$), the transition matrix ($\mathbf{F}$), the selection matrix ($\mathbf{R}$), and the covariance matrix of the state disturbances ($\mathbf{Q}$).

Consider the situation in which we have four dichotomous variables, a single latent factor, a single person, and an observed time series of length $T = 200$. Now, modelling is pursued as follows. We have a four-dimensional vector of observations $\mathbf{y}_t$, a four-dimensional probability vector $\boldsymbol{\pi}_t$, and a four-dimensional linear prediction vector $\boldsymbol{\eta}_t$, related to each other as stated in Equations 7.1 and 7.2. The five-dimensional state vector has the following form

$$\boldsymbol{\alpha}_t = \begin{bmatrix} \theta_t \\ \beta_1 \\ \beta_2 \\ \beta_3 \\ \beta_4 \end{bmatrix}.$$

The specification of the design matrix defines the relation between the person process and the threshold parameters and is given by

$$\mathbf{Z} = \begin{pmatrix} 1 & -1 & 0 & 0 & 0 \\ 1 & 0 & -1 & 0 & 0 \\ 1 & 0 & 0 & -1 & 0 \\ 1 & 0 & 0 & 0 & -1 \end{pmatrix}.$$

The logistic response function relates the linear predictor $\boldsymbol{\eta}_t = \mathbf{Z}\boldsymbol{\alpha}_t$ to the probabilities $\boldsymbol{\pi}_t$

$$\pi_{it} = \frac{\exp(\theta_t - \beta_i)}{1 + \exp(\theta_t - \beta_i)}. \tag{7.3}$$

Equation 7.3 can be seen as a dynamic variant of the Rasch model. Note that this is the form of the Rasch model without the so-called item-invariant discrimination parameter (see Hambleton & Swaminathan, 1985, p. 47). The person process is given by a first-order random walk, that is

$$\theta_t = \theta_{t-1} + \xi_t, \qquad \xi_t \sim \mathrm{N}\,(0, q). \tag{7.4}$$

The random walk can be perceived as the discrete time analogue of Brownian motion (Klebaner, 1998, p. 80). It should be noted that the process in

Equation 7.4 is nonstationary since $\text{Var}(\theta_t) \to \infty$ as $t \to \infty$, and therefore nonergodic. The transition matrix $\mathbf{F}$ is simply the 5×5 identity matrix, $\mathbf{I}_5$, and the selection vector $\mathbf{r}$ is given by

$$\mathbf{r} = \begin{bmatrix} 1 \\ 0 \\ 0 \\ 0 \\ 0 \end{bmatrix}.$$

The initial state vector with associated variance $\mathbf{r}q_0\mathbf{r}'$

$$\mathbf{a}_0 = \begin{bmatrix} \theta_0 \\ \beta_1 \\ \beta_2 \\ \beta_3 \\ \beta_4 \end{bmatrix}.$$

Estimates of the latent process can be obtained with the extended Kalman filter and smoother as described in Fahrmeir (1992). The hyperparameters can be estimated with an EM-type algorithm which is discussed in Fahrmeir and Wagenpfeil (1997). Both procedures are described in the Appendix.

7.5 Analysis of the Dynamic Rasch Model

7.5.1 Simulated Data Example

Data were simulated using the model described in section 7.4.2 with the following hyperparameter settings

$$\mathbf{a}_0 = \begin{pmatrix} \theta_0 \\ \beta_1 \\ \beta_2 \\ \beta_3 \\ \beta_4 \end{pmatrix} = \begin{pmatrix} 0.0 \\ -1.5 \\ -0.5 \\ 0.5 \\ 1.5 \end{pmatrix}, \quad q_0 = 0.1, \quad q = 0.02.$$

For identification purposes, θ_0 is fixed at zero. So we have four items, a single latent factor, a single person, and a series of length $T = 200$. The results of filtered and smoothed probabilities, and filtered and smoothed states are presented. Results were obtained with the following hyperpa-

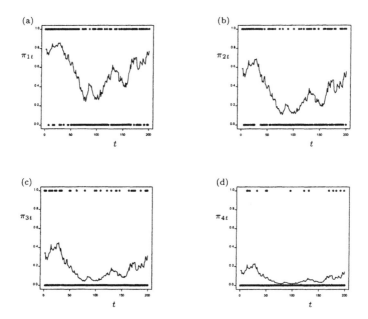

Figure 7.1: True Versus Filtered Probabilities. The Observed Responses are Depicted by Dots.

rameter estimates (standard errors for thresholds could be obtained only)

$$
\hat{a}_0 = \begin{pmatrix} \theta_0 \\ \hat{\beta}_1 \\ \hat{\beta}_2 \\ \hat{\beta}_3 \\ \hat{\beta}_4 \end{pmatrix} = \begin{pmatrix} 0.000 \\ -1.497 & (0.31) \\ -0.846 & (0.30) \\ 0.209 & (0.29) \\ 1.214 & (0.29) \end{pmatrix}, \quad \hat{q}_0 = 0.103, \quad \hat{q} = 0.0197.
$$

The filtering and smoothing results obtained for the probabilities π_{it}, $i = 1, \ldots, 4$, $t = 1, \ldots 200$, are given in Figure 7.1 and Figure 7.2. The solid lines are the true probabilities, the dotted lines are the estimated probabilities, and the dots on the zero- and one-lines are the simulated observations. Figure 7.3 and Figure 7.4 display filtering and smoothing results for the estimation of the latent process. In both figures, the solid line indicates the true latent process and the dotted line the estimated latent process.

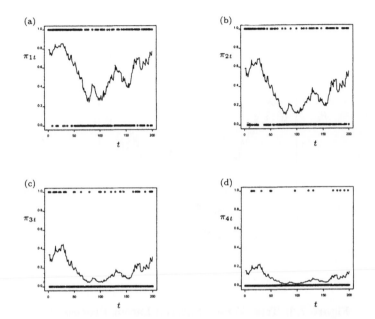

Figure 7.2: True Versus Smoothed Probabilities. The Observed Responses are Depicted by Dots.

Although this is a single simulation, some observations can be made. These are that the filtered and smoothed latent process are somewhat biased, but the direction of the true process is well tracked. The smoother improves the filter estimates considerably, but has some difficulty in reproducing peaks of the true process. Hyperparameter estimates are reasonable, except perhaps the estimates of the threshold parameters.

7.5.2 Real Data Example

Real data were analyzed with the described techniques. We selected a single subject and a single subscale (neuroticism) containing six items of a data set consisting of personality questionnaires containing 30 items scored on seven-point scales, administered to 22 psychology students on 90 consecutive days (Borkenau & Ostendorf, 1998).[5] The questionnaires were constructed as to

---
[5]Data were kindly made available by Dr. Borkenau.

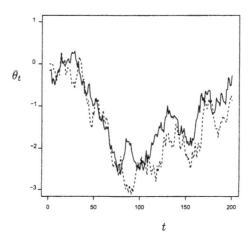

Figure 7.3: True Versus Filtered Latent Process.

measure the big five personality factors. The data were dichotomized for illustrative purposes only to apply the dynamic Rasch model.

Modelling proceeds as in section 7.4. Starting values for the initial state vector $\mathbf{a}_0' = \{\theta_0, \beta_1, \ldots, \beta_6\}$ were obtained by using the logistic transformation of proportions of ones in all items as fixed value for θ_0 and for each single item (β's) corrected for θ_0. Figure 7.5 shows the filtered and smoothed probabilities π_{it}, $i = 1, \ldots, 6$, $t = 1, \ldots 90$, indicated by the solid and dotted lines, respectively. The dots on the zero- and one-lines indicate the dichotomized observations. Results were obtained with the following hyperparameter estimates

$$\hat{\mathbf{a}}_0 = \begin{pmatrix} \theta_0 \\ \hat{\beta}_1 \\ \hat{\beta}_2 \\ \hat{\beta}_3 \\ \hat{\beta}_4 \\ \hat{\beta}_5 \\ \hat{\beta}_6 \end{pmatrix} = \begin{pmatrix} -0.125 \\ -0.216 \ \ (0.52) \\ -0.705 \ \ (0.52) \\ -0.526 \ \ (0.52) \\ 0.565 \ \ (0.52) \\ 0.067 \ \ (0.52) \\ -0.285 \ \ (0.52) \end{pmatrix}, \quad \hat{q}_0 = 0.002, \quad \hat{q} = 0.040.$$

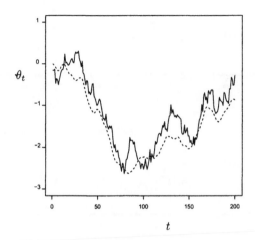

Figure 7.4: True Versus Smoothed Latent Process.

We can see in Figure 7.5 that the filter estimated probabilities vary widely in the beginning of the series. In addition, the difference between the filter and smoother estimates is larger in the first half of the series ($t < 40$). From Figure 7.6, we might infer that this person is relatively stable over time in neuroticism. As we are only concerned with an illustration of the methods, inspection of the fit is not pursued here. However, goodness of fit can be assessed by inspecting deviance statistics, (standardized) residuals or comparing AIC values of competing models (see, e.g., Kedem & Fokianos, 2002, section 1.5).

7.6 Conclusion

In this chapter we took a closer look at the rationale for the emphasis in psychometrics on the analysis of IEV. It was found that this rationale is weak and that arguments for analysis of IAV are too easily brushed aside. We provided arguments for the development of models based on IAV. The question of the existence of any lawful relationship between analysis of IEV

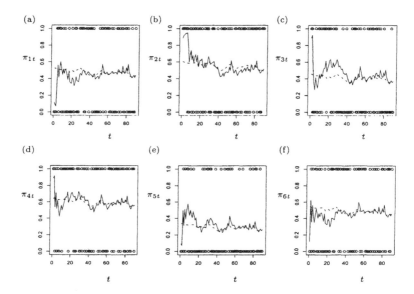

Figure 7.5: Smoothed Versus Filtered Probabilities. The Observed Responses are Depicted by Dots.

and IAV was addressed and it was argued that there are criteria for the existence of such a relationship. These criteria, however, are very strict and are met only when the processes concerned are ergodic. Because, in practice, little is known about the relation between analysis of IEV and IAV, and thus about ergodicity of the processes concerned in psychometrics, investigation of this relation is important. First, however, reliable methods have to be developed for analysis of IAV. This chapter attempted to provide an outline of methods for analyzing single-subject dichotomous time series.

The discussed modelling outline requires further investigation. Models for polytomous responses can be obtained after appropriate adjustments. In addition, it can be investigated if several persons can be analyzed with a model with equal thresholds.

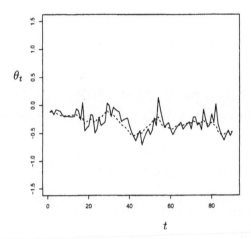

Figure 7.6: Filtered Versus Smoothed Latent Process.

Appendix

Estimation of the Latent Process

The extended Kalman filter and smoother (EKFS) as described in Fahrmeir (1992) are used to obtain estimates of the latent process α_t. In the presentation of the EKFS, the hyperparameters a_0, Q_0, and Q are considered known.

Filtering

First, the filter is initialized by

$$\mathbf{a}_{0|0} = \mathbf{a}_0 \quad \text{and} \quad \mathbf{V}_{0|0} = \mathbf{R}_0 \mathbf{Q}_0 \mathbf{R}'_0.$$

The extended Kalman filter consists of two recursive steps, a prediction and correction step, which are taken consecutively. The prediction step can be described as follows

$$
\begin{aligned}
\mathbf{a}_{t|t-1} &= \mathbf{F}_t \mathbf{a}_{t-1|t-1}, \\
\mathbf{V}_{t|t-1} &= \mathbf{F}_t \mathbf{V}_{t-1|t-1} \mathbf{F}_t' + \mathbf{R}_t \mathbf{Q}_t \mathbf{R}_t'.
\end{aligned}
$$

The correction step is given by the following equations

$$
\begin{aligned}
\mathbf{V}_{t|t} &= \left(\mathbf{V}_{t|t-1}^{-1} + \mathbf{B}_t \right)^{-1}, \\
\mathbf{a}_{t|t} &= \mathbf{a}_{t|t-1} + \mathbf{V}_{t|t} \mathbf{b}_t,
\end{aligned}
$$

where $\mathbf{b}_t$ and $\mathbf{B}_t$ are the score function and the expected information matrix given by

$$
\begin{aligned}
\mathbf{b}_t &= \mathbf{Z}_t' \mathbf{D}_t \boldsymbol{\Sigma}_t^{-1} (\mathbf{y}_t - h(\mathbf{Z}_t \mathbf{a}_{t|t-1})), \\
\mathbf{B}_t &= \mathbf{Z}_t' \mathbf{D}_t \boldsymbol{\Sigma}_t^{-1} \mathbf{D}_t' \mathbf{Z}_t,
\end{aligned}
$$

where $h(.)$ is a logistic response function for our situation and, for the modelling example considered in section 7.4.2

$$
\begin{aligned}
\mathbf{D}_t &= \frac{\partial h(\boldsymbol{\eta}_t)}{\partial \boldsymbol{\eta}_t} \\
&= \begin{pmatrix}
\pi_{t1}(1 - \pi_{t1}) & & & \\
0 & \pi_{t2}(1 - \pi_{t2}) & & \\
0 & 0 & \pi_{t3}(1 - \pi_{t3}) & \\
0 & 0 & 0 & \pi_{t4}(1 - \pi_{t4})
\end{pmatrix},
\end{aligned}
$$

and

$$
\boldsymbol{\Sigma}_t = \begin{pmatrix}
\pi_{t1}(1 - \pi_{t1}) & & & \\
0 & \pi_{t2}(1 - \pi_{t2}) & & \\
0 & 0 & \pi_{t3}(1 - \pi_{t3}) & \\
0 & 0 & 0 & \pi_{t4}(1 - \pi_{t4})
\end{pmatrix}.
$$

Both $\mathbf{D}_t$ and $\boldsymbol{\Sigma}_t$ are symmetric and evaluated at the state prediction $\mathbf{a}_{t|t-1}$. Note that $\mathbf{D}_t$ and $\boldsymbol{\Sigma}_t$ are equal for our modelling example, although in general, this is not the case.

Smoothing

The fixed interval smoother is a backward procedure to obtain smoothed state estimates $\mathbf{a}_{t-1|T}$, which are based on the results of the filtering recursions. For $t = T, \ldots, 1$, we obtain

$$
\begin{aligned}
\mathbf{a}_{t-1|T} &= \mathbf{a}_{t-1|t-1} + \mathbf{G}_t \left(\mathbf{a}_{t|T} - \mathbf{a}_{t|t-1} \right), \\
\mathbf{V}_{t-1|T} &= \mathbf{V}_{t-1|t-1} + \mathbf{G}_t \left(\mathbf{V}_{t|T} - \mathbf{V}_{t|t-1} \right) \mathbf{G}_t',
\end{aligned}
$$

where

$$G_t = V_{t-1|t-1} F_t' V_{t|t-1}^{-1}.$$

Estimation of Hyperparameters

Hyperparameters are estimated with the EM-type algorithm described in Fahrmeir and Wagenpfeil (1997). Only estimation for the time invariant version of the state disturbance covariance matrix is considered. The algorithm for estimating a_0, Q_0, and Q can be described as follows

1. Choose starting values $a_0^{(0)}$, $Q_0^{(0)}$, and $Q^{(0)}$ and set iteration index $p = 0$.

2. Smoothing: Compute $a_{t|T}$, $V_{t|T}$, $t = 1, \ldots, T$ with the above recursions with unknown parameters replaced by their current estimates $a_0^{(p)}$, $Q_0^{(p)}$, and $Q^{(p)}$.

3. EM step: Compute $a_0^{(p+1)}$, $Q_0^{(p+1)}$, and $Q^{(p+1)}$ by

$$
\begin{aligned}
a_0^{(p+1)} &= a_{0|T}^{(p)}, \\
Q_0^{(p+1)} &= V_{0|T}^{(p)}, \\
Q^{(p+1)} &= \frac{1}{T} \sum_{t=1}^{T} \left[\left(a_{t|T}^{(p)} - F a_{t-1|T}^{(p)} \right) \left(a_{t|T}^{(p)} - F a_{t-1|T}^{(p)} \right)' + V_{t|T}^{(p)} \right. \\
&\quad \left. - F G_t^{(p)} V_{t|T}^{(p)} - V_{t|T}^{(p)'} G_t^{(p)'} F' + F V_{t-1|T}^{(p)} F' \right].
\end{aligned}
$$

4. Set $p = p + 1$ and go to 2 until some stopping criterion is reached.

References

Agresti, A. (1997). A model for repeated measurements of a multivariate binary response. *Journal of the American Statistical Association, 92*, 315-321.

Bartholomew, D. J. (1987). *Latent variable models and factor analysis*. New York: Oxford University Press.

Borkenau, P., & Ostendorf, F. (1998). The Big Five as states: How useful is the Five-factor model to describe intraindividual changes over time? *Journal of Research in Personality, 32*, 202-221.

Brillinger, D. R. (1975). *Time series: Data analysis and theory*. New York: Holt, Rinehart, and Winston.

Collins, L. M., & Sayer, A. G. (Eds.). (2001). *New methods for the analysis of change.* Washington, DC: American Psychological Association.

Durbin, J., & Koopman, S. J. (2001). *Time series analysis by state space methods.* New York: Oxford University Press.

Edelman, G. M. (1987). *Neural Darwinism: The theory of neuronal group selection.* New York: Basic Books.

Elliott, R. J., Aggoun, L., & Moore, J. B. (1995). *Hidden Markov models: Estimation and control.* New York: Springer.

Fahrmeir, L. (1992). Posterior mode estimation by extended Kalman filtering for multivariate dynamic generalized linear models. *Journal of the American Statistical Association, 87,* 501-509.

Fahrmeir, L., & Tutz, G. (2001). *Multivariate statistical modelling based on generalized linear models* (2nd ed.). New York: Springer.

Fahrmeir, L., & Wagenpfeil, S. (1997). Penalized likelihood estimation and iterative Kalman smoothing for non-Gaussian dynamic regression models. *Computational Statistics & Data Analysis, 24,* 295-320.

Fischer, G. H. (1983). Logistic latent trait models with linear constraints. *Psychometrika, 48,* 3-26.

Fischer, G. H. (1989). An IRT-based model for dichotomous longitudinal data. *Psychometrika, 54,* 599-624.

Hamaker, E. L., Dolan, C. V., & Molenaar, P. C. M. (2003). ARMA-based SEM when the number of time points T exceeds the number of cases N: Raw data maximum likelihood. *Structural Equation Modeling, 10,* 352-379.

Hambleton, R. K., & Swaminathan, H. (1985). *Item response theory: Principles and applications.* Boston: Kluwer.

Hamilton, J. D. (1994). *Time series analysis.* Princeton, NJ: Princeton University Press.

Harris, C. W. (Ed.). (1962). *Problems in measuring change.* Menasha, WI: The University of Wisconsin Press.

Holtzmann, W. H. (1962). Statistical models for the study of change in the single case. In C. W. Harris (Ed.), *Problems in measuring change* (pp. 199-211). Menasha, WI: The University of Wisconsin Press.

Kedem, B., & Fokianos, K. (2002). *Regression models for time series analysis.* New York: Wiley.

Kelderman, H., & Molenaar, P. C. M. (in press). Comments on the relation between the structure of inter- and intra-individual variability. *Psychometrika.*

Kemeny, J. G., Snell, J. L., & Knapp, A. W. (1966). *Denumerable Markov chains.* Princeton, NJ: Van Nostrand.

Kempf, W. F. (1977). Dynamic models for measurement of "traits" in social behavior. In W. F. Kempf & B. H. Repp (Eds.), *Mathematical models for social psychology* (pp. 14-58). Bern, Switzerland: Hans Huber Publishers.

Klebaner, F. C. (1998). *Introduction to stochastic calculus with applications.* London: Imperial College Press.

Kratochwill, T. R. (Ed.). (1978). *Single subject research: Strategies for evaluating change.* New York: Academic Press.

Lord, F. M., & Novick, M. R. (1968). *Statistical theories of mental test scores.* Reading, MA: Addison-Wesley.

Mackey, M. C. (1992). *Time's arrow: The origins of thermodynamic behavior.* New York: Springer.

Molenaar, P. C. M. (1985). A dynamic factor model for the analysis of multivariate time series. *Psychometrika, 50,* 181-202.

Molenaar, P. C. M., & Raijmakers, M. E. J. (1999). Additional aspects of third source variation for the genetic analysis of human development and behavior. *Twin Research, 2,* 49-52.

Molenaar, P. C. M., & Von Eye, A. (1994). On the arbitrary nature of latent variables. In A. von Eye & C. C. Clogg (Eds.), *Latent variable analysis: Applications for developmental research* (pp. 226-242). Thousand Oaks, CA: Sage.

Molenaar, P. C. M., Boomsma, D. I. & Dolan, C. V. (1993). A third source of developmental differences. *Behavior Genetics, 23,* 519-524.

Molenaar, P. C. M., Huizenga, H. M., & Nesselroade, J. R. (2003). The relationship between the structure of intra-individual and inter-individual variability: A theoretical and empirical vindication of developmental systems theory. In U. M. Staudinger & U. Lindenberger (Eds.), *Understanding human development: Dialogues with lifespan psychology* (pp. 339-360). New York: Kluwer Academic Publishers.

Murray, J. D. (1993). *Mathematical biology* (2nd ed.). New York: Springer.

Nesselroade, J. R., & Schmidt McCollam, K. M. (2000). Putting the process back in developmental processes. *International Journal of Behavioral Development, 24,* 295-300.

Petersen, K. (1983). *Ergodic theory.* Cambridge, England: Cambridge University Press.

Verhelst, N. D., & Glas, C. W. (1993). A dynamic generalization of the Rasch model. *Psychometrika, 58,* 395-415.

Visser, I., Raijmakers, M. E. J., & Molenaar, P. C. M. (2000). Confidence intervals for hidden Markov model parameters. *British Journal of Mathematical and Statistical Psychology, 53,* 317-327.

Chapter 8

The Effect of Missing Data Imputation on Mokken Scale Analysis

L. Andries van der Ark[1] and Klaas Sijtsma
Tilburg University

8.1 Introduction

Tests and questionnaires can be constructed mainly in two ways. The first is exploratory. This means that the final test is selected from the initial set of items so as to optimize psychometric criteria. For example, the test constructor may want to select a subset of items so as to satisfy a lower bound for the reliability of person ordering. The second way of test construction is confirmatory. This means that the set of items is considered to be fixed and the psychometric properties of this set are determined under a particular model without changing the composition of the item set. For example, after fifteen years of use the test constructor may decide that the norms for interpretation of test results need to be updated. The stand-alone software

[1]The first author's research has been supported by the Netherlands Research Council (NWO), Grant No. 400.20.011. Thanks are due to Liesbeth van den Munckhof for her assistance with the MSP analyses and Joost van Ginkel for correcting an error in the initial computation of the statistic MIN.

package MSP (Molenaar & Sijtsma, 2000) allows both possibilities. A well known problem in data analysis for test and questionnaire construction is that some of the N respondents did not supply an answer to some of the J items, so that the data matrix $\mathbf{X}$ is incomplete. MSP only offers listwise deletion to handle the missing data problem. This may result in the loss of many cases, biased estimates of parameters of interest, and reduced accuracy of estimates. The topic of this chapter is the comparison of imputation methods with respect to the outcomes of exploratory and confirmatory test construction as implemented in MSP.

8.1.1   Missing Data Mechanisms

Missing item scores may be due to many reasons. Often these reasons are unknown to the researcher. For example, the respondent may have missed a particular item (e.g., due to inattention or time pressure), missed a whole page of items, saved the item for later and then forgot about it, did not know the answer and then left it open, became bored while taking the test or questionnaire and skipped a few items, felt the item was embarrassing (e.g., questions about one's sexual habits), threatening (questions about the relationship with one's children), or intrusive to privacy (questions about one's income and consumer habits), or felt otherwise uneasy and reluctant to answer.

Rubin (1976; also, see Little & Rubin, 1987; Schafer, 1997) formalized mechanisms of missing data into three classes. Let i denote the respondent index and j the item index, and let x_{ij} be the integer score of respondent i on item j. Let $\mathbf{M}$ be an $N \times J$ indicator matrix of with elements $m_{ij} = 1$ if score x_{ij} is missing, and $m_{ij} = 0$ if score x_{ij} is observed. The observed part of $\mathbf{X}$ is denoted $\mathbf{X}_{obs}$ and the missing part is denoted $\mathbf{X}_{mis}$. Thus, $\mathbf{X} = (\mathbf{X}_{obs}, \mathbf{X}_{mis})$. Let β be a set of parameters governing the data, $\mathbf{X}_{obs}$ and $\mathbf{X}_{mis}$, and ξ a set of parameters governing the missingness, $\mathbf{M}$. We may model the distribution of the missing data as $P(\mathbf{M}|\mathbf{X}_{mis}, \mathbf{X}_{obs}, \beta, \xi)$.

The missing data are called *missing at random* (MAR) when the distribution of the missing data does not depend on the missing item scores; that is

$$P(\mathbf{M}|\mathbf{X}_{mis}, \mathbf{X}_{obs}, \beta, \xi) = P(\mathbf{M}|\mathbf{X}_{obs}, \xi)P(\mathbf{X}_{obs}|\beta).$$

An example of MAR is that missing item scores depend on other observed items or covariates. Such a covariate may be gender. For example, for men it may be more difficult to admit to the item 'I cry at weddings' than for women (item taken from questionnaire by Vingerhoets & Cornelius, 2001). Therefore, a larger proportion of the male respondents may decide not to respond to this item.

A special case of MAR is *missing completely at random* (MCAR). Data are MCAR when the missing data values are a simple random sample of all data values; that is,

$$P(\mathbf{M}|\mathbf{X}_{mis}, \mathbf{X}_{obs}, \beta, \xi) = P(\mathbf{M}|\xi).$$

For MCAR the parameters in ξ only affect the proportion of missing values, but not the pattern of missingness.

Missing data are called *nonignorable* when their distribution $P(\mathbf{M}|\mathbf{X}_{mis}, \mathbf{X}_{obs}, \beta, \xi)$ depends on $\mathbf{X}_{obs}$, $\mathbf{X}_{mis}$, and ξ, and indirectly on β since these parameters govern $\mathbf{X}_{obs}$ and $\mathbf{X}_{mis}$. One example of a nonignorable missingness mechanism is that the distribution of the missing data depends on values of variables that were not part of the investigation. For example, in a personality inventory missingness may depend on general intelligence or reading ability. Another example of a nonignorable missingness mechanism is that the distribution of the missing data depends on the missing item scores; for example, respondents who cry at weddings have a higher probability of not answering the item 'I cry at weddings' than respondents who never cry at weddings. Consequently, any missing data method based on available item scores would underestimate the missing value.

8.1.2 Test Construction

Exploratory and confirmatory test construction

Our frame of reference in this study is nonparametric item response theory (NIRT; Boomsma, Van Duijn, & Snijders, 2001; Mokken, 1971; Sijtsma & Molenaar, 2002; Van der Linden & Hambleton, 1997). Following NIRT, we define a latent trait θ that stands for a psychological property or a collection of psychological properties measured by the J items. For example, the item "I cry at weddings" may be indicative of the latent trait "tendency to cry". Parameter θ thus governs the data and replaces parameter vector β. Let X_j be the random variable for the score on item j. Item scores may be dichotomous or polytomous. For example, the item "I cry at weddings" may have only two answer categories, "applies" and "does not apply", which may be dichotomously scored $x_j = 1$ and $x_j = 0$ with respect to latent trait "tendency to cry", respectively. Another possibility is that the respondent indicates on an ordered rating scale the degree to which the item applies to him/her, and the corresponding polytomous scoring then may be $x_j = 0, \ldots, g$. Latent trait θ is estimated by means of $X_+ = \sum_j X_j$ (Hemker, Van der Ark, & Sijtsma, 2001; Junker, 1991; Stout, 1990). Note that X_+ may either estimate a unidimensional θ or a multidimensional θ.

The construction of a test or questionnaire mainly follows two possibilities. The first possibility is that one starts from scratch, defining the construct of interest and a useful operationalization, and then defines a collection of experimental items. Then a clustering method from MSP may be used to determine the structure of the data in terms of the underlying latent traits. A cluster is a set of items that measure the same latent trait. This is an exploratory approach because the dimensionality structure was not hypothesized prior to the application of the clustering method but found by the program. The second possibility is that one starts with an existing instrument and wants to know whether it can be used in another population or at a later point in time. This entails drawing a new sample of respondents to which the existing item set is administered, or administering the item set to the same respondents once more. Then MSP may be used to analyze the item set as one cluster and determine its psychometric properties. Because the item set is considered to be fixed, we consider this kind of item analysis to be confirmatory in the sense that for this set it is determined whether or not it is a useful instrument in a new context.

Test construction according to MSP

Scalability coefficients.   Both for exploratory and confirmatory test construction, MSP uses the scalability coefficient H (Mokken, 1971, pp. 148-153; 1997; Sijtsma & Molenaar, 2002, pp. 49-64) as a scaling criterion. For two items j and k, $Cov(X_j, X_k)$ defines their covariance and $Cov(X_j, X_k)_{\max}$ defines their maximum covariance given the marginal distributions of their bivariate frequency table. The scalability coefficient for these two items is defined as

$$H_{jk} = \frac{Cov(X_j, X_k)}{Cov(X_j, X_k)_{\max}}$$

Coefficient H_{jk} is the basis for the scalability coefficient of one item with respect to the other $J - 1$ items; this coefficient is denoted H_j and defined as

$$H_j = \frac{\sum\limits_{k \neq j}^{J} Cov(X_j, X_k)}{\sum\limits_{k \neq j}^{J} Cov(X_j, X_k)_{\max}}$$

Finally, scalability coefficient H for all J items is defined as

$$H = \frac{\sum_{j=1}^{J-1} \sum_{k=j+1}^{J} Cov(X_j, X_k)}{\sum_{j=1}^{J-1} \sum_{k=j+1}^{J} Cov(X_j, X_k)_{\max}}$$

Monotone homogeneity model. The use of scalability coefficients H_{jk}, H_j, and H is related to the monotone homogeneity model (MHM; Mokken, 1971, p. 118). The MHM assumes a unidimensional latent trait θ, local independence of the item scores given θ, and a monotone nondecreasing relationship between $P(X_j \geq x_j|\theta)$ and θ. For scores $x_j = 1, \ldots, g$, the conditional probabilities $P(X_j \geq x_j|\theta)$ are the item step response functions (ISRFs) (for $x_j = 0$ the ISRF equals 1 by definition). For dichotomous items $(g = 1)$ the only relevant ISRF is $P(X_j \geq 1|\theta) = P(X_j = 1|\theta)$. This is the item response function (IRF). Together, the assumptions of unidimensionality, local independence, and monotonicity define the MHM. For dichotomous items, the MHM implies the stochastic ordering of latent trait θ by means of observable summary score X_+; that is, for any t, we have that $P(\theta > t|X_+)$ is nondecreasing in X_+ (based on Grayson, 1988; also, see Hemker, Sijtsma, Molenaar, & Junker, 1997). Thus, the MHM implies ordinal person measurement on θ using X_+. The more complicated case for polytomous items is treated by Van der Ark (in press).

Relationship between MHM and coefficient H. The MHM implies that $H_{jk} \geq 0$ (Holland & Rosenbaum, 1986; Mokken, 1971, pp. 149-150). By implication, we have that $H_j \geq 0$ and $H \geq 0$. Based on these implications, Mokken (1971, p. 184; Sijtsma & Molenaar, 2002, pp. 67-68) defined a scale as a set of dichotomously scored items for which, for a suitably chosen positive constant c, and for product-moment correlation ρ,

$$\rho_{jk} > 0, \text{ for all item pairs } (j, k); \tag{8.1}$$

and

$$H_j \geq c > 0, \text{ for all items } j. \tag{8.2}$$

Equation 8.1 implies that $H_{jk} > 0$. Equation 8.1 also implies that $H_j > 0$ and $H > 0$. In addition, by specifying that $H_j \geq c$, Equation 8.2 poses minimum requirements on the slope of the IRF. That is, constant c forces a minimum level of discrimination power on the individual items. This is not implied by the MHM, but because this model allows weakly sloped IRFs and even flat IRFs as a borderline case, the addition of a minimum discrimination requirement is a practical measure for reliable person ordering. Finally, the definition of a scale can be extended readily to polytomous items (Sijtsma & Molenaar, 2002, p. 127).

Automated item selection. For exploratory test construction, MSP selects items according to the definition of a scale (Equations 8.1 and 8.2). The default option for item selection, to be used here, has the following steps (Mokken, 1971, pp. 190-194).

1. From the J available items, MSP selects from the item pairs which have a H_{jk} that is significantly greater than 0, that pair which has the highest H_{jk} that is greater than c. This is the start set for item selection.

2. From the remaining $J - 2$ items, that item is added to the start set that (a) has a positive covariance with both selected items (Equation 8.1); (b) has an H_j value with the selected items that is at least c (Equation 8.2); and (c) has the highest common H value with the selected items, given all candidate items for selection.

3. The next items are selected following the logic of Step 2. The item selection for the first scale ends when no more items satisfy the criteria mentioned in Step 2.

4. If items remain unselected after the first scale has been formed, from the unselected items MSP tries to form a second scale, a third scale, and so on, until no more items remain or no more items satisfy the criterion in Step 1.

For confirmatory test construction, the MHM is fitted to the data corresponding to the a priori defined test consisting of J items using methods implemented in MSP (Molenaar & Sijtsma, 2000; Sijtsma & Molenaar, 2002). This includes calculating and evaluating the H_j and H coefficients.

8.2   Methods for Missing Data Imputation

We introduce four methods for the imputation of item scores for missing observations in a data matrix $\mathbf{X}$, plus listwise deletion. Listwise deletion is the only method currently implemented in MSP. It was used as a benchmark for the other methods. For each of the five methods it was investigated how they influence the results of the automated item selection procedure in MSP (exploratory test construction) and how they influence the results of fitting the MHM to an a priori defined scale (confirmatory test construction). The five missing data handling methods are discussed next.

Listwise Deletion. Listwise deletion (LD) deletes from the analysis all cases that have at least one missing item score. Because for data matrices that contained at least ten percent missing item scores it was found that

LD led to the rejection of almost the whole data matrix, in these cases we used the imputation of a random item score as an alternative (called *Random Imputation*; abbreviated RI).

Two-Way Imputation. Because in a unidimensional test or questionnaire all item scores measure the same latent trait, the scores on the available items can be used for imputing scores for missing data. Let PM_i be the mean item score of person i calculated across his/her available item scores; let IM_j be the mean score on item j calculated across the item scores available in the sample of N persons; and let OM be the mean item score calculated across all available item scores in **X**. Then for missing item score (i, j), we calculate

$$TW_{ij} = PM_i + IM_j - OM; \ TW_{ij} \in \mathbb{R}.$$

The item score to be imputed is obtained by rounding TW_{ij} to the nearest feasible integer. Two-way imputation (TW) was proposed by Bernaards and Sijtsma (2000; see Huisman & Molenaar, 2001, for a related method).

Response Function Imputation. Response function imputation (RF; Sijtsma & Van der Ark, 2003) is based on the idea to impute item scores x_{ij} as random draws from the distribution $P(X_j = x_j | \theta_i)$. The steps in this procedure are the following.

- First, estimate θ_i by means of restscore $R_{i(-j)} = X_{i+} - X_{ij}$ (e.g., Hemker, et al., 1997; Junker, 1993; Sijtsma & Molenaar, 2002, p. 40). This is done as follows. Due to missing data, the number of available item scores on the remaining $J - 1$ items may vary across respondents. This number is denoted J_i ($J_i \leq J - 1$). Restscore $R_{i(-j)}$ is computed as the sum of these available item scores. Because different respondents may have different numbers of available item scores, to have all restscores on the same scale each restscore is multiplied by $(J - 1)/J_i$.

- Second, estimate $P(X_j = x_j | \theta_i)$ by means of $P[X_j = x_j | R_{i(-j)}]$, for $x_j = 0, \ldots, m$. The latter probability is computed in the subgroup having an observed score on X_j. Each respondent's X_j is weighted by the accuracy with which his/her restscore, $R_{i(-j)}$, estimates its expectation, $E_i[R_{i(-j)}]$. Because for each respondent one restscore is available, the determination of its accuracy is based on its constituent J_i item scores. Let the mean item score of respondent i be denoted $\overline{X}_i = \frac{R_{i(-j)}}{J_i}$. Let σ_i^2 denote the variance of the item scores of respondent i, estimated by $S_i^2 = \frac{\sum_j (X_{ij} - \overline{X}_i)^2}{J_i}$. The inaccuracy of $\overline{X}_i$ is given by $SE(\overline{X}_i) = \sqrt{S_i^2 / J_i}$. The weight for respondent i in computing $P[X_j = x_j | R_{i(-j)}]$ is $1/SE(\overline{X}_i)$.

- Third, for a missing score in cell (i, j) we impute a random draw from $P[X_j|R_{i(-j)}]$. In the subgroup of people having a missing score on item j, restscores may exist that did not exist in the group with X_j observed that was used for estimating $P[X_j|R_{i(-j)}]$. For example, among the latter group $R_{i(-j)} = 2$ may not have been observed; thus, $P[X_j|R_{i(-j)} = 2]$ was not estimated. In that case, item score probabilities are obtained by linear interpolation between the two nearest restscores from the group with X_j observed. If restscore groups are too small for an accurate estimate of $P[X_j|R_{i(-j)}]$, adjacent restscore groups may be joined. See Sijtsma and Van der Ark (2003) for more details.

Multiple Response Function Imputation. Multiple response function imputation (MRF) entails five times the application of the RF procedure. This involves five random draws from $P[X_j|R_{i(-j)}]$, which yields five different completed data matrices. Each completed data matrix is analyzed separately, and the results are combined later using Rubin's rules (see, e.g., Schafer, 1997, pp. 109-110) or a variation to be discussed later.

Multiple multivariate normal imputation. An imputation method for categorical data proposed by Schafer (1997, pp. 257-275) and implemented in publicly available software (program CAT; Schafer, 1998a) was considered for item score imputation. This method requires a frequency table based on J items with $m+1$ answer categories, which thus has $(m+1)^J$ entries. In our applications, this number was too large for maximum likelihood estimation of the imputation model. Thus, CAT could not be used. Instead we assumed a multivariate normal imputation model as suggested by Schafer (1997, p. 148; program NORM, Schafer, 1998b). The method is called multiple multivariate normal imputation (MMNI). Method MMNI assumes that the item scores have a J-variate normal distribution. In an initial step the model parameters, the mean vector and the covariance matrix, are estimated using an EM algorithm. Then an iterative procedure called *data augmentation* is used to obtain the distribution of the missing item scores given the observed item scores and the model parameters. The missing values are imputed by random draws from this conditional distribution. Since these random draws are real-valued and our data integer-valued, the random draws were rounded to the nearest feasible integer. For more detailed information on data augmentation we refer to Tanner and Wong (1987) and for the implementation of EM and data augmentation in NORM to Schafer (1997, chap. 5 and chap. 6).

8.3   Method

We investigated the influence of each of the five imputation methods on the results of confirmatory and exploratory item analysis using the program MSP. Three real data sets (first design factor) were used. These data sets are referred to as *original* data sets.

- **Verbal analogies data** (Meijer, Sijtsma, & Smid, 1990). For this data set, $N = 990$ and $J = 32$, with $g + 1 = 2$. This test measures verbal intelligence in adults. Meijer et al. (1990) found that 31 items together formed one scale (each $H_j > 0$). This was the basis for the confirmatory analysis. All 32 items were used in the exploratory analysis.

- **Coping data** (Cavalini, 1992). For this data set, $N = 828$ and $J = 17$, with $g + 1 = 4$. This questionnaire measures coping styles in response to industrial malodors. Cavalini (1992, pp. 53-54) found four item subsets (17 items in total) measuring different coping styles. Each of these subsets was used separately in the confirmatory analysis. The set of 17 items was the input for the exploratory analysis.

- **Crying data** (Vingerhoets & Cornelius, 2001). Here, $N = 3965$ and $J = 54$, with $g + 1 = 7$. This questionnaire measures determinants of adult crying behavior. Scheirs and Sijtsma (2001) found three subsets of items (54 items in total), representing three psychological states. Each subset was the basis of the confirmatory analysis. All 54 items together were subjected to the exploratory analysis.

Each data set was complete. In each original data set item scores were deleted using procedures that resulted in either MCAR, MAR, or nonignorable missingness (second design factor). The percentage of missing item scores was either 5%, 10%, or 20% (third design factor). The data sets containing missing data are referred to as *incomplete* data sets. Missingness was simulated as follows

- **MCAR.** The probability of a missing score was the same for each entry in the data set.

- **MAR.** Let $L = \mathrm{trunc}(J/2)$ be a cut-off value that splits the item set into a first half (items $1, \ldots, L$) and a second half (items $L+1, \ldots, J$). When the missing item scores were MAR, the probability of a missing item score in the second half was twice the probability of a missing item score in the first half.

- **Nonignorable missingness.** When missingness was nonignorable, the missing item scores were MAR in combination with the following mechanism: Let $G = \text{trunc}(g/2)$ be a cut-off value that splits the item scores into low item scores $(0, \ldots, G)$ and high item scores $(G + 1, \ldots, g)$. The probability of a missing value for high item scores was twice the probability of a missing value for low item scores.

The incomplete data sets were imputed using *listwise deletion* (5% missing item scores) or *random imputation* (10% and 20% missing item scores), *two-way imputation*, *response function imputation*, *multiple response function imputation*, and *multiple multivariate normal imputation* (fourth design factor). These data sets are referred to as *completed* data sets. Both the original and the completed data sets were subjected to exploratory and confirmatory data analysis (fifth design factor).

Exploratory analysis. For the single imputation methods (RI, TW, and RF), for each incomplete data set, the MCAR, MAR, and nonignorable missingness conditions were used to construct three different completed data sets. For each completed data set, MSP found a cluster solution, which was compared with the original data cluster solution. Assume that an item set consists of five items, indexed $j = 1, \ldots, 5$, then the original-data clustering might be $(1, 2, 2, 0, 1)$: The 1 scores indicate that items 1 and 5 were in the same cluster, the 2 scores that items 2 and 3 were in another cluster, and the 0 score that item 4 remained unselected. Now, assume that the completed-data clustering is $(1, 1, 1, 0, 0)$; then, ignoring the cluster numbering (which is nominal) the smallest number of items to be moved to reobtain the original-data solution is sought. Here, items 1 and 5 need to be moved to a separate cluster. Denote the minimum number of items to be moved by MIN (with realization min), then for this example $MIN = 2$.

For the multiple imputation methods (MRF and MMNI), for each incomplete data set five completed data sets were generated. The five completed-data cluster solutions were combined to one by taking the mode of the cluster indices for each item. For example, let the five cluster solutions found be $(1, 2, 2, 0, 1)$, $(2, 2, 1, 0, 1)$, $(1, 2, 1, 1, 2)$, $(1, 2, 2, 0, 1)$, and $(0, 2, 2, 0, 0)$; then, the modal solution is $(1, 2, 2, 0, 1)$ and the MIN value with respect to the original-data clustering, which was $(1, 2, 2, 0, 1)$ (previous example), is determined to be 0.

Confirmatory analysis. The H values of the completed data were compared with the H values of the corresponding original data. For multiple imputation the mean H of the five completed data matrices was taken.

The design was completely crossed with 3 (original data matrices) × 3 (missingness mechanisms) × 3 (percentages of missing item scores) × 5

Table 8.1: Number of Verbal Analogies Items Incorrectly Clustered in Exploratory Analysis, for Five Imputation Methods, Three Missingness Mechanisms, and Three Percentages (5, 10, and 20) of Imputed Item Scores $[J = 32; \max(MIN) = 18]$.

Method	Missingness Mechanism								
	MCAR			MAR			Nonignorable		
	5	10	20	5	10	20	5	10	20
LD/ RI	13	18	18	10	18	16	8	18	18
TW	8	14	16	5	15	16	4	9	16
RF	4	3	8	5	3	7	3	8	4
MRF	2	2	7	5	6	9	3	3	4
MMNI	10	17	17	12	11	16	6	12	16

(imputation methods) × 2 (exploratory vs. confirmatory analysis) = 270 cells. The study was programmed in S-Plus 6 for Windows (2001); the exploratory and confirmatory analyses were done using MSP (Molenaar & Sijtsma, 2000).

8.4   Results

8.4.1   Exploratory Analyses

Table 8.1 (Verbal Analogies data), Table 8.2 (Coping data), and Table 8.3 (Crying data) give the value of MIN for the complete design. An unscalable set of items is one in which each item forms a unique cluster; for this setup MIN was determined, and the result was called $\max(MIN)$. The value of $\max(MIN)$ was used as a benchmark.

Verbal analogies data. Methods LD and RI always led to almost one half to all items incorrectly clustered ($8 \leq min \leq 18$). Method TW led to a misclassification of almost all items for 10% and 20% imputed item scores. Methods RF and MRF performed best ($2 \leq min \leq 8$). Method MMNI led to high MIN-values ($6 \leq min \leq 17$). This result was not expected and may be related to convergence to a local optimum. This is further elaborated in the Discussion.

Coping data. For 5% imputed item scores, all methods performed well. For 10% and 20% imputed item scores, method RI led to large values of MIN. Methods TW, RF, and MRF led to the misclassification of approximately one-fifth of the items for 10% imputed item scores, and to

Table 8.2: **Number of Coping Data Items Incorrectly Clustered in Exploratory Analysis, for Five Imputation Methods, Three Missingness Mechanisms, and Three Percentages (5, 10, and 20) of Imputed Item Scores** $[J = 17; \max(MIN) = 12]$.

Method	Missingness Mechanism								
	MCAR			MAR			Nonignorable		
	5	10	20	5	10	20	5	10	20
LD/RI	1	6	10	1	6	10	1	7	10
TW	0	3	6	1	3	5	0	1	4
RF	0	2	6	0	2	5	0	4	4
MRF	0	1	6	0	2	4	0	3	5
MMNI	0	0	0	0	0	1	0	0	0

Table 8.3: **Number of Crying Data Items Incorrectly Clustered in Exploratory Analysis, for Five Imputation Methods, Three Missingness Mechanisms, and Three Percentages (5, 10, and 20) of Imputed Item Scores** $[J = 54; \max(MIN) = 45]$.

Method	Missingness Mechanism								
	MCAR			MAR			Nonignorable		
	5	10	20	5	10	20	5	10	20
LD/RI	10	16	29	9	17	34	11	21	38
TW	5	3	10	2	7	5	3	3	12
RF	5	4	7	2	4	6	3	6	10
MRF	3	4	6	3	5	7	1	6	10
MMNI	21	16	44	25	36	44	16	32	44

Table 8.4: Bias in H (in hundredths; i.e., -2 stands for $-.02$) for One Cluster of Verbal Analogies Items, for Five Imputation Methods, Three Missingness Mechanisms, and Three Percentages (5, 10, and 20) of Imputed Item Scores ($J = 31, H = .25$).

Method	Missingness Mechanism								
	MCAR			MAR			Nonignorable		
	5	10	20	5	10	20	5	10	20
LD/ RI	1	9	5	-1	-16	-20	0	-15	-19
TW	-3	-5	-9	-2	-9	-10	-2	-4	-7
RF	0	0	-1	0	0	-1	0	0	-1
MRF	0	0	-1	0	0	-1	0	0	-1
MMNI	-2	-5	-10	-2	-7	-10	-2	-5	-9

the misclassification of approximately one-third of the items for 20% imputed item scores. Method MMNI led to a correct clustering except for 20% item scores that were MAR. Only small differences were found among the missing data mechanisms MCAR, MAR and nonignorable.

Crying data. Method LD/RI led to a misclassification of approximately one-fifth (5% missing item scores, $min = 9$) to two-thirds (20% missing item scores, $min = 38$) of the items. Method MMNI resulted in even higher MIN-values ($16 \leq min \leq 44$). Similar to the results for the Verbal Analogies data (Table 8.1), this is probably due to a bad model-fit. Methods TW, RF, and MRF performed best and yielded misclassifications of approximately one-tenth (5% and 10% imputed item scores) to one-fifth (20% imputed item scores) of the items. Only small differences were found among the missing data mechanisms MCAR, MAR and nonignorable.

8.4.2 Confirmatory Analysis

Table 8.4 (Verbal Analogies data), Table 8.5 (Coping data), and Table 8.6 (Crying data) give the bias in H for the entire design of a single predefined cluster of a data set. The bias is defined as H of the completed data minus H of the original data. For notational convenience the fractional divisions and leading zeros are omitted. Thus, a bias notation of -2 stands for -0.02.

Verbal analogies data. For 5% imputed item scores all imputation methods led to a small bias (Table 8.4). For 10% and 20% imputed item scores, methods TW and MMNI led to a negative bias between $-.10$ and

Table 8.5: Bias in H (in hundredths; i.e., -2 stands for $-.02$) for Four Clusters of Coping Data Items, for Five Imputation Methods, Three Missingness Mechanisms, and Three Percentages (5, 10, and 20) of Imputed Item Scores (Cluster I: $J = 7$, $H = .31$; Cluster II: $J = 4$, $H = .50$; Cluster III: $J = 3$, $H = .56$; Cluster IV: $J = 3$, $H = .35$).

Method	MCAR			MAR			Nonignorable		
	5	10	20	5	10	20	5	10	20
	Cluster I								
LD/RI	1	−7	−17	−1	−10	−16	−2	−9	−17
TW	0	0	2	1	0	0	0	1	2
RF	1	0	−2	0	−1	−3	−1	0	−3
MRF	0	0	−2	0	−1	−3	0	−1	−2
MMNI	0	1	−1	0	0	0	0	0	−1
	Cluster II								
LD/RI	−1	−18	−27	−2	−20	−29	1	−16	−31
TW	−1	−6	−2	−3	−7	−7	−2	−6	−7
RF	0	−3	−6	−2	−2	−11	−2	−3	−7
MRF	−1	−3	−7	−2	−4	−10	−1	−4	−9
MMNI	1	−2	−1	0	−1	−1	0	−2	−3
	Cluster III								
LD/RI	−2	−13	−21	1	−9	−13	−2	−8	−16
TW	1	3	3	1	1	2	1	1	4
RF	−2	−4	−14	0	−1	−3	−1	−1	−5
MRF	−2	−3	−13	0	−1	−3	−1	−1	−3
MMNI	−2	−2	−1	0	0	−2	−1	0	−2
	Cluster IV								
LD/RI	1	−9	−14	2	−9	−14	0	−9	−16
TW	3	4	7	4	6	13	3	9	16
RF	0	−2	−1	2	−3	−5	−3	−2	−3
MRF	0	−2	−3	0	−3	−4	1	−2	−6
MMNI	0	−1	0	1	0	1	1	1	3

Table 8.6: Bias in H (in hundredths; i.e., -2 stands for $-.02$) for Three Clusters of Crying Data Items, for Five Imputation Methods, Three Missingness Mechanisms, and Three Percentages (5, 10, and 20) of Imputed Item Scores (Cluster I: $J = 22$, $H = .43$; Cluster II: $J = 14$, $H = .41$; Cluster III: $J = 18$, $H = .30$).

| Method | Missingness Mechanism | | | | | | | | |
| | MCAR | | | MAR | | | Nonignorable | | |
	5	10	20	5	10	20	5	10	20
	Cluster I								
LD/ RI	1	-12	-20	0	-13	-22	-2	-12	-22
TW	-1	-2	-4	-1	-2	-4	-2	-4	-6
RF	-1	-1	-3	0	-1	-3	-1	-2	-5
MRF	-1	-1	-3	-1	-1	-3	-1	-2	-5
MMNI	0	0	0	0	-1	0	0	-1	-1
	Cluster II								
LD/ RI	-1	-9	-16	2	-9	-16	0	-10	-17
TW	-2	-4	-7	-2	-4	-7	-3	-6	-9
RF	0	-1	-2	0	-1	-2	-1	-1	-4
MRF	0	0	-2	0	-1	-2	0	-1	-4
MMNI	0	0	0	0	0	0	0	0	-1
	Cluster III								
LD/ RI	0	-10	-17	0	-10	-16	-1	-10	-16
TW	0	0	-1	0	0	-1	0	-1	-1
RF	-1	-1	-3	-1	-1	-3	-1	-2	-4
MRF	-1	-1	-4	-1	-1	-3	-1	-1	-4
MMNI	0	0	-1	0	0	-1	0	-1	-1

−.04. Methods RF and MRF performed best yielding unbiased or almost unbiased results in all cases.

Coping data. The results for the four clusters of the Coping data are presented in Table 8.5. For Cluster I, all methods except LD/RI yielded a small bias in H in all conditions; method MMNI gave the best results.

For Cluster II, method LD/RI had a small bias for 5% missing item scores and a large negative bias for 10% and 20% missing item scores. Methods TW, RF, and MRF had a small negative bias within the range $[-.07, .00]$, for 5% and 10% imputed item scores, and a larger negative bias within the range $[-.11, -.02]$, for 20% imputed item scores. Method MMNI was the most successful method, the largest bias in H being −.03.

Similar to Cluster II, for Cluster III method LD/RI showed a large negative bias for 10% and 20% imputed item scores. Method TW led to a small positive bias in H, and method MMNI led to a small negative bias. Methods RF and MRF showed a large negative bias $(-.14)$ in H when applied to data with 20% item scores that were MCAR. This unexpected result may be related to the small number of items in Cluster III. This is further elaborated in the Discussion.

Similar to Cluster II and Cluster III, for Cluster IV method LD/RI showed a large negative bias for 10% and 20% imputed item scores. Methods RF, MRF, and MMNI gave the best bias results, which were between −.06 and .03. Method TW showed large positive bias (.07, .13, and .16) when applied to data with 20% imputed item scores. This unexpected result may also be related to the small number of items in Cluster IV.

For all item clusters it was found that there were only small differences among MCAR, MAR, and nonignorable missingness. It was also found for all clusters that methods RF and MRF produced approximately the same results.

Crying data. The results for the three clusters of the Crying data are presented in Table 8.6. The results were similar for the three clusters. For 5% imputed item scores all methods led to a small bias in H within the range $[-.03, 02]$. For 10% and 20% imputed item scores, methods TW, RF, MRF, and MMNI produced satisfactory results although, when applied to Cluster II, method TW produced a bias that was a little higher (within the range $[-.04, -.09]$). Method MMNI performed best. There were only small differences among MCAR, MAR, and nonignorable missingness.

8.5   Discussion

This chapter showed that using method LD in Mokken scale analysis can result in cluster solutions that deviate much from the cluster solutions that would have been obtained had the data been complete. For 10% and 20% missingness, the number of cases left may be so small that Mokken scale analysis becomes impossible. These results are in line with earlier studies on method LD (e.g., Schafer, 1997, p. 23). The alternative benchmark, method RI, led to large values of MIN and large biases in H.

By using total scores on the J items, methods TW, RF, and MRF make use of the property that all items are indicators of the same latent variable. The advantage of method TW is its simplicity, which makes the method easy to use for researchers. The values of MIN and the bias in H resulting from method TW were large for the Verbal Analogies data and smaller for the Coping data and the Crying data.

The results for methods RF and MRF were similar. The main reason for choosing multiple imputation over single imputation is to obtain more stable results and correct standard errors. For Mokken scale analysis the standard errors of H usually do not play an important role, and the bias and the values of H produced by methods RF and MRF were similar. Thus, we could not demonstrate the advantage of method MRF over method RF. Methods RF and MRF are not as simple as method TW and involve some computational decisions, such as the sample size of the restscore-groups and the weight given to each restscore. In general, methods RF and MRF performed a little better than method TW with respect to MIN values and bias.

We found a large bias in H for imputation methods RF and MRF, for a cluster of 3 items (Coping data, Cluster III), 20% missingness, and missingness mechanism MCAR. When $J = 3$, the restscore is based on two items. Given these conditions, theoretically under MCAR it is expected that 32% of the sample has a missing score on one item and 4% of the sample has missing scores on both items. This may have caused inaccurate rest-score estimates which led to the large bias.

Method MMNI yielded the lowest MIN-values and the smallest bias of all methods when the number of items was less than 23 (Crying data, Cluster I). For larger item sets (Verbal Analogies data $[J = 31]$, and the Crying data $[J = 54]$), the results for method MMNI were worse than the results for method LD/RI. The reason may be the EM-algorithm in program NORM reached a local optimum for which the fit was much worse than the required fit. The algorithm then kept iterating (without improvement) until the maximum number of iterations was reached, yielding a badly fitting model. Consulting the auxiliary statistics provided by NORM and keeping track of the number of iterations may prevent the researcher from using these wrong estimates. The successor of NORM, which is incorporated in the software package S-plus 6 for Windows (2001), gives an error message in these situations without supplying completed data.

Currently, a more systematic investigation (Van Ginkel, Van der Ark, & Sijtsma, 2004) is conducted to determine the effect of multiple imputation using the methods discussed here on results of Mokken scaling and several other psychometric methods. Using simulated data, several comprehensive designs were analyzed to obtain a more definitive impression about the usefulness of our (multiple) imputation methods.

References

Bernaards, C. A., & Sijtsma, K. (2000). Influence of imputation and EM methods on factor analysis when item nonresponse in questionnaire data is nonignorable. *Multivariate Behavioral Research, 35,* 321-364.

Boomsma, A., Van Duijn, M. A. J., & Snijders, T. A. B. (Eds.) (2001). *Essays on item response theory.* New York: Springer.

Cavalini, P. M. (1992). *It's an ill wind that brings no good: Studies on odour annoyance and the dispersion of odour concentrations from industries.* Unpublished doctoral dissertation. University of Groningen, The Netherlands.

Grayson, D. A. (1988). Two-group classification in latent trait theory: Scores with monotone likelihood ratio. *Psychometrika, 53,* 383-392.

Hemker, B. T., Sijtsma, K., & Molenaar, I. W., & Junker, B. W. (1997). Stochastic ordering using the latent trait and the sum score in polytomous IRT models. *Psychometrika, 62,* 331-347.

Hemker, B. T., Van der Ark, L. A., & Sijtsma, K. (2001). On measurement properties of continuation ratio models. *Psychometrika, 66,* 487-506.

Holland, P. W., & Rosenbaum, P. R. (1986). Conditional association and unidimensionality in monotone latent variable models. *The Annals of Statistics, 14,* 1523-1543.

Huisman, J. M. E., & Molenaar, I. W. (2001). Imputation of missing scale data with item response models. In A. Boomsma, M. A. J. van Duijn & T. A. B. Snijders (Eds.), *Essays on item response theory* (pp. 221-244). New York: Springer.

Junker, B. W. (1991). Essential independence and likelihood-based ability estimations for polytomous items. *Psychometrika, 56,* 255-278.

Junker, B. W. (1993). Conditional association, essential independence, and monotone unidimensional item response models. *The Annals of Statistics, 21,* 1359-1378.

Little, R. J. A., & Rubin, D. B. (1987). *Statistical analysis with missing data.* New York: Wiley.

Meijer, R. R., Sijtsma, K., & Smid, N. G. (1990). Theoretical and empirical comparison of the Mokken and the Rasch approach to IRT. *Applied Psychological Measurement, 14*, 283-298.

Mokken, R. J. (1971). *A theory and procedure of scale analysis.* The Hague: Mouton/Berlin: De Gruyter.

Mokken, R. J. (1997). Nonparametric models for dichotomous responses. In W. J. Van der Linden & R. K. Hambleton (Eds.), *Handbook of modern item response theory* (pp. 352-367). New York: Springer

Molenaar, I. W., & Sijtsma, K. (2000). *User's manual MSP5 for Windows.* Groningen, The Netherlands: iecProGAMMA.

Rubin, D. B. (1976). Inference and missing data. *Biometrika, 63*, 581-592.

Schafer, J. L. (1997). *Analysis of incomplete multivariate data.* London: Chapman & Hall.

Schafer, J. L. (1998a). CAT. Software for S-PLUS Version 4.0 for Windows. Retrieved from *http://www.stat.psu.edu/~jls/sp40.html.*

Schafer, J. L. (1998b). NORM. Software for S-PLUS Version 4.0 for Windows. Retrieved from *http://www.stat.psu.edu/~jls/sp40.html.*

Scheirs, J. G. M., & Sijtsma, K. (2001). The study of crying: Some methodological considerations and a comparison of methods for analyzing questionnaires. In A. J. J. M. Vingerhoets & R. R. Cornelius (Eds.), *Adult Crying. A Biopsychosocial Approach* (pp. 279-298). Hove, UK: Brunner-Routledge.

Sijtsma, K., & Molenaar, I. W. (2002). *Introduction to nonparametric item response theory.* Thousand Oaks, CA: Sage.

Sijtsma, K., & Van der Ark, L. A. (2003). Investigation and treatment of missing item scores in test and questionnaire data. *Multivariate Behavioral Research, 38*, 505-528.

S-Plus 6 for Windows. [Computer software.] (2001). Seattle, WA: Insightful Corporation.

Stout, W. F. (1990). A new item response theory modelling approach with applications to unidimensionality assessment and ability estimation. *Psychometrika, 55*, 293-325.

Tanner, M. A., & Wong, W. H. (1987). The calculation of posterior distributions by data augmentation (with discussion). *Journal of the American Statistical Association, 82*, 528-550.

Van der Ark, L. A. (in press). Stochastic ordering of the latent trait by the sum score under various polytomous IRT models. *Psychometrika.*

Van der Linden, W. J., & Hambleton, R. K. (Eds.) (1997). *Handbook of modern item response theory.* New York: Springer.

Van Ginkel, J. R., Van der Ark, L. A., & Sijtsma, K. (2004). *Multiple imputation of item scores in test and questionnaire data, and influence on psychometric results.* Manuscript submitted for publication.

Vingerhoets, A. J. J. M., & Cornelius, R. R. (Eds.) (2001). *Adult Crying: A Biopsychosocial Approach.* Hove, UK: Brunner-Routledge.

Chapter 9

Building IRT Models From Scratch: Graphical Models, Exchangeability, Marginal Freedom, Scale Types, and Latent Traits

Henk Kelderman
Vrije Universiteit Amsterdam

9.1 Introduction

Because measurements in the Behavioral and Social Sciences are often quite fallible, tests consist of a number of separate items whose responses are combined to obtain a more reliable test score (Cronbach, 1951). This, of course, only makes sense if all items measure the same attribute, which implies that their response probabilities should satisfy certain requirements embodied in a, say, common attribute criterion (CAC). To date, a CAC is formulated as a statistical model that explicitly contains a latent variable (CACL). The latent variable, say θ, represents the degree to which a subject possesses the common attribute. See Hambleton and Van der Linden (1997) for an overview. Because the interpretation of the latent variable is independent of a particular test, they are easily interpreted as the 'true'

167

attribute. Consequently they are very attractive from a realist perspective (Borsboom, 2003, p. 49). However some conceptual problems arise.

Because θ is unobserved, the precise nature of its relations with the item responses and other variables cannot be determined on empirical grounds. So, one needs a theoretical justification. However, unlike Physics, where theory is almost always cast in mathematical form, the Behavioral and Social Sciences rarely have such theories and are of a purely semantic nature (Borsboom, 2003, p. 5). At best, these theories make statements about the absence or presence of relations in a nomological net (Cronbach & Meehl, 1955; Hempel, 1965). As a result, the definition of the nature of these relations is usually based on tradition, mathematical convenience, or even faith.

Because in many research applications, the focus is on the existence of relations of the attribute with other constructs in the nomological net, it is usually considered not necessary to use latent variables. Unfortunately, in practice, most researchers use a rather crude score function such as the simple sum of arbitrarily scored item responses. Obviously, these scoring functions are far from optimal because items may vary widely in quality and meaning. Consequently, even if it is not necessary or one is not willing to explicitly assume a latent variable underlying the test response, one still has to have an appropriate measurement model that yields a suitable test score under some defensible CAC.

This chapter starts by proposing some minimal qualitative criteria that the joint distribution of item responses and other variables in the nomological net should obey if they are to measure a common attribute. We formulate two CACs in terms of manifest variables only (CACMs). Although all these criteria are justified by the fact that the items should measure a common attribute they are nonparametric in the sense that no latent variable is explicitly introduced to describe the distribution. To a certain extend, these assumptions provide a justification for parametric latent variable models.

Next we describe CACMs embodying various relaxations of DeFinetti exchangeability of measurements, depending on what part of the distribution is of interest in a particular application. Various exchangeability models for nominal data are discussed and represented as graphical models. It is shown that under certain realistic assumptions exchangeability models may also be formulated for metric item scores.

To set the stage, we start with CACMs, based on properties of the independence graph of the observed variables.

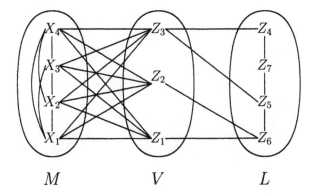

Figure 9.1: Independence Graph for Exchangeable Measures in a Nomological Net: X_1-X_4, **Items of Raven's Progressive Matrix Test;** Z_1, **Peripheral Nerve Conduction;** Z_2, **Velocity Frequency of the Alpha Wave;** Z_3, **Cerebral Glucose Metabolism;** Z_4, **Impulsiveness;** Z_5, **Crime Prevalence;** Z_6, **School Success;** Z_7, **Job Success.**

9.2 Graphical Criteria

CACMs based on graphical criteria are only concerned with the structure of relations of the item responses and other variables in the *independence graph* (Cox & Wermuth; 1996). Figure 9.1 depicts an undirected independence graph G_O describing the (in)dependence structure of a set of observed variables. The *vertices* in the graph denote variables $X_1, \ldots, X_4$, $Z_1, \ldots, Z_7$. If two variables are connected (not-connected) by an *edge* they are dependent (independent) given the remaining variables in the graph. M, V and L are subsets of variables, where M contains items that are supposed to measure a common attribute, V is the set of collateral variables to which the responses M are directly related, and L are variables to which they are not directly related. Because in most studies subjects respond independently from each other to categorically scored items, we assume that the joint distribution is *Poisson, multinomial,* or *product multinomial,* depending on which aspects of the distribution are of interest and depending on the sampling scheme under which the data were generated. Furthermore we assume, for simplicity, that all item responses have the same number of response categories k.

As a hypothetical example, consider a set M of appropriate measures

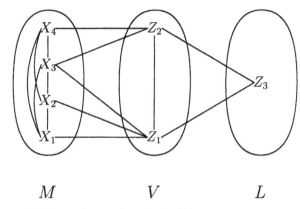

$$M \qquad\qquad V \qquad\qquad L$$

Figure 9.2: Independence Graph for Exchangeable Measures in a Nomological Net: X_1-X_2, **Orientation in Time;** X_3-X_4, **Orientation in Space;** Z_1, **Age;** Z_2, **Dementia as Diagnosed From Neurological Measurements;** Z_3, **Gender.**

of general intelligence such as the items of Raven's progressive matrix test (Raven, Raven, & Court, 1991), a set V that contains the physiological measures peripheral nerve conduction, Z_1, velocity frequency of the alpha wave, Z_2, and cerebral glucose metabolism, Z_3, and a set L with measures of impulsiveness, Z_4, crime prevalence, Z_5, school success, Z_6, and job success, Z_7 (Jensen, 1998, chap. 6).

The only statistical hypothesis that an undirected graphical model states is that (sets of) variables that are not connected by an edge are independent given the variables that indirectly connect them in the graph. For example, in Figure 9.1 the variables X_1, a test item, and Z_7, job success, are independent given the physiological variables Z_1, Z_2, and Z_3.

As another example consider a set of items constructed to measure dementia. In Figure 9.2 M contains four items, where the first two items measure orientation in time, and the last two items orientation in space. The other variables from the nomological net are age, Z_1, dementia as diagnosed from neurological measurements, Z_2, and gender, Z_3. It is seen that the item responses are independent of gender given age and dementia measured by neurological instruments.

The graphs in Figure 9.1 and Figure 9.2 are undirected and do not contain any information about the causal direction of the relations. Of course there are pervasive substantive arguments that some of the relations in the graphs are causal, for example, the relation between cerebral glucose metabolism and measured intelligence and between age and dementia. However, because we want to focus on the properties of the item responses and keep the discussion of CACMs completely general, we limit ourselves at this stage to the undirected graph.

9.2.1 Qualitative Internal Consistency

The qualitative internal consistency (QIC) CACM only imposes restrictions on the configuration of the edges between the item responses. Its definition is as follows.

Definition 1 (qualitative internal consistency) *The set of measures M is a clique.*

A clique is a graph-theoretical term for a maximally connected subgraph, that is, a subgraph where all members of the clique are connected to all other members of that subgraph. It is seen in Figure 9.1 and Figure 9.2 that in both examples the items satisfy the QIC CACM.

The rationale of this CACM is as follows. If two items are conditionally independent in the graph, there are two possibilities: Either the items measure different attributes or the items measure an attribute that is perfectly measured by the remaining variables in the graph. In the first case one or both items should be removed, in the second case both items would be superfluous and should be removed. So a set of test items should satisfy QIC.

9.2.2 Qualitative External Consistency

The next criterion for items to measure same attribute and that attribute alone is qualitative external consistency (QEC). It imposes restrictions on the configuration of the edges between the elements of M and the remaining vertices in the graph. The *boundary*, bd(a), of a set a is the set of vertices that have an edge with one or more edges in a. In Figure 9.1 and Figure 9.2 the set V is the boundary of M. The vertices on the boundary of the set of test items may be called collateral variables. The QEC criterion states that all items should have the same set of collateral variables. QEC can formally be defined as

Definition 2 (qualitative external consistency)

$$\bigcap_{i \in M} \text{bd}(i) = \text{bd}(M).$$

It is readily verified that M in the graph in Figure 9.1 satisfies the QEC criterion, but the graph in Figure 9.2 does not. In the dementia example, the items measuring orientation in time have a direct relation to age. That relation is missing in the items measuring orientation in space. Thus, elderly people may not necessarily need to suffer from dementia, Z_2, to lose track of time. They do, however, usually know where they are located, unless they are demented. If items concerning orientation in time have different collateral variables than items concerning orientation in space, both types of items cannot measure the same attribute and that attribute alone. Which items are functioning differentially and which are not depends on what one wants the test to measure.

If one item of the progressive matrix test is directly related with school success and the others are not, some subjects may have had an opportunity to learn the solution process of that particular item but not that of the other items. If general intelligence is an attribute not directly related to education, the item functions differentially. Conversely, if general intelligence is directly related to cerebral glucose metabolism, every single measure of general intelligence should be directly related to it. If an item does not, it fails to measure the intended attribute and also functions differentially.

It should be noted that if a set of items satisfies QIC and QEC, it does not necessarily mean that all items measure the same attribute. That depends to a large extend on whether the graph is sufficiently complete to represent the nomological network around the attributes measured by the test. For example, the union of two subsets of items measuring two different attributes that have the same sets of collateral variables in the graph may satisfy QEC. In that case the graph may represent too small a part of the nomological net to distinguish both attributes. As another example consider a graph where there are no collateral variables and where each item measures a different attribute. If the attributes are all positively related, such as is the case with intelligence factors, the total set of items may well satisfy QIC.

QIC and QEC are purely qualitative CACMs. They depend only on the presence or absence of edges in the independence graph. They are independent in the sense that one does not imply the other, although empirically a violation of the QEC may go very well together with a violation of QIC and vice versa. QIC and QEC are qualitative in the sense that they do not say anything about the strength of the relations between the variables in the graph. If it is deemed unacceptable that some items are more strongly related to some variables in V than other items, one must impose additional restrictions on the items' relations. One way to do this, is to require that the items' relations are in some sense exchangeable.

9.3 Exchangeability

Because the variables in M are all independent of L given V, we may limit ourselves to describing the joint distribution of $M \cup V$. The strongest CAC can be derived from DeFinetti's (1937) notion of exchangeability applied to the item responses.

The idea is as follows. If it is completely indifferent which of the items is used, their responses should have perfectly identical statistical properties in the nomological net. In that case, the joint distribution should be permutation invariant in the item responses, that is, the item responses should be exchangeable in the sense of DeFinetti (1937).

Denote item responses that are in some sense exchangeable by $X_i, i = 1, \ldots, m$, where m is the number of items, that is, the number of elements, $|M|$, of M. Let the coordinate projections $\mathbf{X} = (X_i; i \in M)$ and $\mathbf{Z} = (Z_j; j \in V)$ denote random vectors with realizations $\mathbf{x}$ and $\mathbf{z}$ and with joint distribution $P(\mathbf{Z}, \mathbf{X})$.

DeFinetti measurement exchangeability (DME) is formally defined as

Definition 3 (DeFinetti Measurement Exchangeability) *For all permutations* $\mathbf{x}^* = \mathrm{perm}(\mathbf{x})$

$$P(\mathbf{Z} = \mathbf{z}, \mathbf{X} = \mathbf{x}) = P(\mathbf{Z} = \mathbf{z}, \mathbf{X} = \mathbf{x}^*). \tag{9.1}$$

Note the following properties of DME. First, this condition is not affected by a single one-to-one transformation of all x_i ($i \in M$). Second, DME implies that any coefficient describing the association between item responses x_i and z_j ($j \in V$) must, by symmetry, be invariant over items and that the associations between responses are all the same. Third, DME also holds for all subsets a of M because integrating out unpermuted X_i, $i \in M - a$, does not change DME.

Gulliksen (1968) hinted at this CAC when he required that measurements of the same attribute should be *interchangeable* in the sense that 'it is indifferent as to which of the measures is used.' Lazarsfeld (1959, pp. 113-117) proposed a similar CAC (see also Mokken, 1971, p. 2). Huynh (1978), and more recently Schuster (2001), applied the DME CAC to formulate a model for raters to determine the extend to which it is indifferent which of the raters is used.

However, the DME CACM seems quite unrealistic for most applications because, by symmetry, it implies that the distributions $P(\mathbf{x}_a)$ for all subsets $a \subset M$ should be identical, that is

$$P(\mathbf{X}_a = \mathbf{x}_a) = P(\mathbf{X}_{a'} = \mathbf{x}_a), \ |a| = |a'|. \tag{9.2}$$

which, for $|a| = 1$, implies

$$P(X_i = x_i) = P(X_{i'} = x_i). \tag{9.3}$$

In most practical measurement situations, marginal distributions of the item responses may be different due to systematic or random measurement errors. In principle, DME can be made insensitive to them by ignoring certain response marginals.

9.3.1 Ignoring Marginal Distributions

DME's implication (Equation 9.2) can be removed by conditioning out the distribution of the item responses (Equation 9.1). Joint-variable conditioning measurement exchangeability (JCME) is then defined as

Definition 4 (Joint-variable Conditioning Measurement Exchangeability) *For all permutations* $\mathbf{x}^* = \text{perm}(\mathbf{x})$,

$$P(\mathbf{Z} = \mathbf{z}|\mathbf{X} = \mathbf{x}) = P(\mathbf{Z} = \mathbf{z}|\mathbf{X} = \mathbf{x}^*). \tag{9.4}$$

It is easily shown that Equation 9.4 together with Equation 9.2 is equivalent to DME.

JCME ignores the joint marginal distribution of the item responses. Therefore, the CAC is only sensitive to the associations with collateral variables V. It is insensitive to deviations from DME due to systematic measurement errors affecting associations within M. These measurement errors may be caused by method factors that are not necessarily related to the collateral variables. Test-taking artifacts may occur when applying measures in a short time after one another (Spearman, 1910).

Instead of leaving the joint distribution of the item responses unrestricted in DME, one may relax each of the single-variable marginal item response distributions $P(x_i)$ in DME. This yields the single-variable conditioning measurement exchangeability (SCME) CAC defined formally as

Definition 5 (Single-variable Conditioning Measurement Exchangeability) *For all* $a \subseteq M$

$$P(\mathbf{Z} = \mathbf{z}, \mathbf{X}_b = \mathbf{x}_b|\mathbf{X}_a = \mathbf{x}_a) = P(\mathbf{Z} = \mathbf{z}, \mathbf{X}_b = \mathbf{x}_b|\mathbf{X}_a = \mathbf{x}_a^*), \tag{9.5}$$

where $b = M - a$.

It is readily shown that Equation 9.5 and Equation 9.3 are jointly equivalent to Equation 9.1. SCME may be defended on the grounds that random or systematic method factors that affect only a single item may level each other out in the test score and would only lead to a uniform lowering of the relations with other variables in the nomological net. In the next section the log-linear model is used to test CME.

9.4 Log-linear Models For Exchangeability

A general parametric model for categorical data is the log-linear model (LLM). A LLM arises by setting $\log P(\mathbf{xz})$ equal to a linear model. In this chapter, we take the linear model to be a hierarchical model with main effects describing the effects of one variable and interaction effects describing the effects of the combination of two or more variables. To keep equations short, we denote the sum of all elementary main and interaction terms within a set of variables by the term of the highest order. It is often instructive to specify one or more lower order terms explicitly. In that case it assumed that the term is subtracted from the sum of terms that is specified in the model. For example, the sum of all main and interaction effects of the set of measures a is denoted by $\lambda_{\mathbf{x}_a}^{\mathbf{X}_a}$ whereas in $\lambda_{x_i}^{X_i} + \lambda_{\mathbf{x}_a}^{\mathbf{X}_a}$, with $i \in a$, $\lambda_{\mathbf{x}_a}^{\mathbf{X}_a}$ denotes the sum of all main and interaction effects except $\lambda_{x_i}^{X_i}$. Thus, the fully saturated hierarchical LLM can be written as

$$\log P(\mathbf{xz}) = \lambda + \lambda_{\mathbf{x}} + \lambda_{\mathbf{z}} + \lambda_{\mathbf{xz}}, \qquad (9.6)$$

or alternatively

$$P(\mathbf{xz}) = \exp(\lambda + \lambda_{\mathbf{x}} + \lambda_{\mathbf{z}} + \lambda_{\mathbf{xz}}),$$

where superscripts are suppressed for simplicity. In Model 9.6, λ is the general mean effect, $\lambda_{\mathbf{x}}$ the sum of main and interaction effects of and between $\mathbf{x}$, $\lambda_{\mathbf{z}}$ the sum of main and interaction effects of and between $\mathbf{z}$, and $\lambda_{\mathbf{xz}}$ the sum of interaction effects between $\mathbf{x}$ and $\mathbf{z}$. To obtain an identifiable model, the elementary λ-parameters are constrained to sum to zero over each scalar index.

To model DME one has to set

$$\lambda_{\mathbf{x}} = \lambda_{\mathbf{x}^*} \quad \text{and} \quad \lambda_{\mathbf{xz}} = \lambda_{\mathbf{x}^* \mathbf{z}}. \qquad (9.7)$$

An alternative way to model exchangeability is to use sum scores. Let $t_{ih}(x_i) = I(x_i, h)$, where $I(x_i, h) = 1$ if $x_i = h$ and 0 otherwise, and let $\mathbf{t}_i(x_i) = (t_{i1}(x_i), \ldots, t_{ih}(x_i), \ldots, t_{ik}(x_i))$. Note that $\mathbf{t}_i(x_i)$ is a one-to-one function of x_i. Furthermore, let

$$\mathbf{t}(\mathbf{x}) = \sum_i \mathbf{t}_i(x_i).$$

As addition is associative, we have

$$\mathbf{t}(\mathbf{x}) = \mathbf{t}(\mathbf{x}^*),$$

so instead of the restriction in Equation 9.7, we may set

$$\lambda_{\mathbf{x}} = \lambda_{\mathbf{t}} \quad \text{and} \quad \lambda_{\mathbf{xz}} = \lambda_{\mathbf{tz}}$$

to model DME.

To model JCME we do not put restrictions on the parameter corresponding to the main and interaction effects of the item responses $\lambda_{\mathbf{x}}$, so we have

$$\log P(\mathbf{xz}) = \lambda + \lambda_{\mathbf{x}} + \lambda_{\mathbf{z}} + \lambda_{\mathbf{tz}}. \tag{9.8}$$

Note that this model does not contain a term $\lambda_{\mathbf{t}}$ because $\mathbf{t}$ is a function of $\mathbf{x}$ so that $\lambda_{\mathbf{t}}$ can be fully absorbed in $\lambda_{\mathbf{x}}$. The model satisfies JCME because it is easily shown that because there is a one-to-one correspondence of $\lambda_{\mathbf{x}}$ and $P(\mathbf{x})$, the marginal distribution $P(\mathbf{x})$ is unrestricted.

The model for SCME is

$$\log P(\mathbf{xz}) = \lambda + \sum_i \lambda_{x_i} + \lambda_{\mathbf{z}} + \lambda_{\mathbf{t}} + \lambda_{\mathbf{tz}}. \tag{9.9}$$

In this model, the term $\lambda_{\mathbf{t}}$ should be present because it describes exchangeable associations between the item responses, whereas the λ_{x_i} describe only main effects. Furthermore, it is easily shown that in this model, the marginal distributions $P(x_i)$ are unrestricted. Thus the model corresponds to SCME.

Graphical models for SCME

Klein, Keiding, and Kreiner (1995) studied the graphical version of SCME extensively. In SCME the interactions between the item responses and between collateral variables and item responses can all be described by their relations with $\mathbf{t}$. Therefore, the undirected graph G_T, Figure 9.3, has a vertex representing the composite test score random variable $\mathbf{T}$. However, G_T is a quasi-independence graph rather than an independence graph; since $\mathbf{T}$ is a function of $\mathbf{X}$, the probability space of $(\mathbf{X}', \mathbf{T}', \mathbf{Z}')'$ is restricted (an apostrophe denotes the transpose of the column vector). Haberman (1979) describes the theory of log-linear models for restricted probability spaces. In this theory, variables are *quasi-independent* if—according to the model—they would be independent in the unrestricted probability space. A quasi-independence model eliminates the associations due to structural constraints on the probability space from the empirical associations. Because graphical models for discrete data are essentially systems of log-linear models, we can also formulate graphical models for restricted probability spaces. The graph then becomes a quasi-independence graph rather than an independence graph. The corresponding graphical model is developed as follows:

Let $\Omega_{\mathbf{X}}$, $\Omega_{\mathbf{T}}$, and $\Omega_{\mathbf{Z}}$ denote the probability spaces of $\mathbf{X}$, $\mathbf{T}$, and $\mathbf{Z}$ respectively. Let Ξ be the probability space of $(\mathbf{X}', \mathbf{T}')'$, that is, the subset

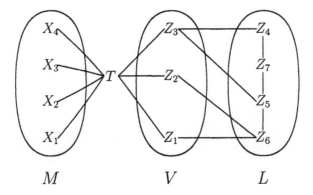

Figure 9.3: **Quasi-Independence Graph for Exchangeable Measures and Test Scores in a Nomological Net:** X_1-X_4, **Items of Raven's Progressive Matrix Test;** Z_1, **Peripheral Nerve Conduction;** Z_2, **Velocity Frequency of the Alpha Wave;** Z_3, **Cerebral Glucose Metabolism;** Z_4, **Impulsiveness;** Z_5, **Crime Prevalence;** Z_6, **School Success;** Z_7, **Job Success.**

of the Carthesian product $\Omega_\mathbf{X} \times \Omega_\mathbf{T}$ that satisfies the functional constraints. Consequently, the corresponding probabilities $P(\mathbf{xtz})$ satisfy

$$P(\mathbf{xtz}) = P(\mathbf{xz}), \quad (\mathbf{x}',\mathbf{t}',\mathbf{z}')' \in \Xi \times \Omega_\mathbf{Z}, \quad (\mathbf{x}',\mathbf{z}')' \in \Omega_\mathbf{X} \times \Omega_\mathbf{Z}.$$

Using this equation, the quasi-independence graph can now be defined in terms of the parametric log-linear models for measurement exchangeability

$$P(\mathbf{xtz}) = P(\mathbf{x}|\mathbf{t})P(\mathbf{t}|\mathbf{z})P(\mathbf{z}), \qquad (9.10)$$

where each (conditional) distribution in the right hand side is described by the appropriate (quasi) log-linear model for the probability spaces $(\mathbf{x}',\mathbf{t}')' \in \Xi$, $(\mathbf{z}',\mathbf{t}')' \in \Omega_\mathbf{Z} \times \Omega_\mathbf{T}$, and $\mathbf{t} \in \Omega_\mathbf{T}$, respectively.

Similarly, for SCME one has conditional quasi independence of the item responses given the score

$$P(\mathbf{x}|\mathbf{t}) = \prod_i P(x_i|\mathbf{t}), \quad (\mathbf{x}',\mathbf{t}')' \in \Xi,$$

because of the additivity in

$$P(\mathbf{x}|\mathbf{t}) \propto \exp\left(\sum_i \lambda_{x_i}\right), \quad (\mathbf{x}',\mathbf{t}')' \in \Xi. \qquad (9.11)$$

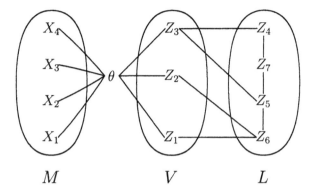

Figure 9.4: Independence Graph for Exchangeable Measures With Latent Variable in a Nomological Net: Intelligence Example.

The corresponding quasi-independence graph G_T is depicted in Figure 9.3. Thus, SCME in G_0 is equivalent to quasi independence in G_T. It is seen from Figure 9.3 that in G_T all empirical associations among X_1-X_4,Z_1-Z_3 in the restricted probability space are explained by T.

9.5   Latent Variables

In latent variable models, it is assumed that the measures are uniquely related to a vector of latent variables $\boldsymbol{\theta}$, which means that given $\boldsymbol{\theta}$, each item response is independent of all other variables, that is

$$P(\mathbf{xz\theta}) = \prod_i P(x_i|\boldsymbol{\theta})P(\boldsymbol{\theta}|\mathbf{z})P(\mathbf{z}).$$

The corresponding independence graph G_L is depicted in Figure 9.4.

A general SCME-type LLM for the associations between item response $\mathbf{x}$ and latent variables $\boldsymbol{\theta}$ is

$$P(x_i|\boldsymbol{\theta}) = c_i(\boldsymbol{\theta})^{-1} \exp \left[\lambda_{x_i} + \sum_h t_{ih}(x_i)\theta_h \right].$$

The proportionality constant $c_i(\boldsymbol{\theta})$ ensures that the probabilities sum to one. Assuming independence of the x_i given $\boldsymbol{\theta}$, the model for the joint response is

$$P(\mathbf{x}|\boldsymbol{\theta}) = c(\boldsymbol{\theta})^{-1} \exp\left(\sum_i \lambda_{x_i} + \mathbf{t}'\boldsymbol{\theta}\right). \tag{9.12}$$

From this we have

$$P(\mathbf{t}|\boldsymbol{\theta}) = c(\boldsymbol{\theta})^{-1} \exp(\mathbf{t}'\boldsymbol{\theta})\gamma(\mathbf{t}), \tag{9.13}$$

with

$$\gamma(\mathbf{t}) = \sum_{\mathbf{x}}(\exp \sum_i \lambda_{x_i}), \quad (\mathbf{x}', \mathbf{t}')' \in \Xi.$$

Dividing Equation 9.12 by Equation 9.13 we have

$$P(\mathbf{x}|\mathbf{t}\boldsymbol{\theta}) = \frac{P(\mathbf{xt}|\boldsymbol{\theta})}{P(\mathbf{t}|\boldsymbol{\theta})} = \gamma(\mathbf{t})^{-1} \exp\left(\sum_i \lambda_{x_i}\right), \quad (\mathbf{x}', \mathbf{t}')' \in \Xi,$$

which does not depend on $\boldsymbol{\theta}$. So we have

$$P(\mathbf{x}|\mathbf{t}\boldsymbol{\theta}) = P(\mathbf{x}|\mathbf{t}) \propto \exp\left(\sum_i \lambda_{x_i}\right), \quad (\mathbf{x}', \mathbf{t}')' \in \Xi,$$

which is identical to Equation 9.11.

Because $\mathbf{t}$ is a function of $\mathbf{x}$ we can write

$$P(\mathbf{x}|\boldsymbol{\theta}) = P(\mathbf{xt}|\boldsymbol{\theta}) = P(\mathbf{x}|\mathbf{t}\boldsymbol{\theta})P(\mathbf{t}|\boldsymbol{\theta}) = P(\mathbf{x}|\mathbf{t})P(\mathbf{t}|\boldsymbol{\theta}),$$

so one obtains

$$P(\mathbf{xz}\boldsymbol{\theta}) = P(\mathbf{x}|\mathbf{t})P(\mathbf{t}|\boldsymbol{\theta})P(\boldsymbol{\theta}|\mathbf{z})P(\mathbf{z}).$$

The model for the observed data is then

$$P(\mathbf{xz}) = P(\mathbf{x}|\mathbf{t})P(\mathbf{t}|\mathbf{z})P(\mathbf{z}), \tag{9.14}$$

where

$$P(\mathbf{t}|\mathbf{z}) = \int_{\boldsymbol{\theta}} P(\mathbf{t}|\boldsymbol{\theta})P(\boldsymbol{\theta}|\mathbf{z})d\boldsymbol{\theta}. \tag{9.15}$$

Because by Equation 9.15, $P(\mathbf{t}|\mathbf{z})$ is restricted to be consistent with an underlying latent variable $\boldsymbol{\theta}$, CACL (Equation 9.14), is a special case of the CACM (Equation 9.10). For the case of the dichotomous Rasch model Cressie and Holland (1984) and Hout, Duncan, and Sobel (1987) have studied the consequences this restriction. The restriction leads to a complicated set of inequality constraints on the moments of the test score distribution. These constraints are violated if there are gross differences in the probabilities of consecutive score values. So to be consistent with Equation 9.15 the test score distribution should be smooth.

The class of log-linear latent variable models contains a number of well-known logistic item response models. Relevant for SCME is Rasch' model for polytomous items (Rasch, 1960; Andersen, 1973). This model assumes that the propensity of answering in a particular type of category h can be described by a latent variable θ_h $(h = 0, \ldots, k)$. Let δ_{x_i} be a parameter describing the difficulty of response x_i. To make the parameters identifiable $\delta_0 = 0$ and $\theta_0 = 0$. The vector of subject parameters is denoted by $\boldsymbol{\theta} = (\theta_1, \ldots, \theta_h, \ldots, \theta_k)'$. The polytomous Rasch model for the probability that a subject with latent values $\boldsymbol{\theta}$ gives a response x_i is

$$P(x_i|\boldsymbol{\theta}) = c_i\left(\boldsymbol{\theta}\right)^{-1} \exp\left(\sum_{h=1}^{k} t_{ih}\theta_h - \delta_{x_i}\right).$$

The conditional probability of the joint response is

$$P(\mathbf{x}|\boldsymbol{\theta}) = c(\boldsymbol{\theta})^{-1} \exp\left(\sum_{h=1}^{k} t_h\theta_h - \sum_i \delta_{x_i}\right),$$

which is equivalent to Model 9.12 if $\sum_{h=1}^{k} t_h\theta_h$ is written as the vector product $\mathbf{t}'\boldsymbol{\theta}$ and δ_{x_i} as $2\left(\lambda_0^{X_i} - \lambda_{x_i}^{X_i}\right)$, to obtain a contrast with the zero category.

9.6 Some Generalizations and Restrictions

The models so far all assume that item responses are purely qualitative and that the category scores x_i only serve to identify a certain response category. Furthermore, it is assumed that the x_i $(i = 1, \ldots, m)$ are all the same for categories with the same meaning and different for categories with different meaning. This assumption makes the models quite restrictive since they imply that all items are equally good predictors of their collateral variables.

One obvious way to generalize CME models is to use category scores with metric properties. Kelderman (in press) studied exchangeability of *transformed* continuous multivariate normally distributed item responses. Similarly, for multinomially distributed categorical variables one could transform each nominal category score x_i into a metric category score $u_i = s_i(x_i)$, where s_i is a one-to-one function with as domain the set of natural numbers and range the set of real numbers. Using this transformation it seems that one can obtain exchangeability models for metrically scored item responses. The only formal differences between metric exchangeability models and nominal exchangeability models is then that the metric category scores u_i must be exchangeable rather than the nominal scores x_i and $t_{ih}(x_i) = I(x_i, h)$ must be replaced by $t_{ih}(x_i) = I(x_i, h)u_i$ in all LLMs.

However, by allowing the category scores to be real numbers, serious difficulties arise. As the metric category scores u_i are real numbers they are unique, so that the sum score

$$t_h(\mathbf{x}) = \sum_i I(x_i, h)u_i$$

identifies the items for which the response is equal to h. Consequently, **t** will be exactly equivalent to the vector of original responses **x** (Mokken, 1971, p. 103), so that $\lambda_{\mathbf{t}}$ is equivalent to $\lambda_{\mathbf{x}}$ and $\lambda_{\mathbf{tz}}$ equivalent to $\lambda_{\mathbf{xz}}$. Therefore, in the metric case, all versions of exchangeability LLMs discussed so far degenerate into the saturated LLM (Equation 9.6). To make the models nondegenerate one must put constraints on the parameters $\lambda_{\mathbf{t}}$ and/or $\lambda_{\mathbf{tz}}$.

The first way to make CME models nondegenerate is to make use of the metric property of u_i to restrict the interaction parameters $\lambda_{\mathbf{tz}}$ in G_T to conform to an association model. For simplicity consider the case of one collateral variable z_j and some dichotomous items scored $u_i = 0, 1$. So one has $x_i = t_{i1} = 0, 1$ and one must set $t_0 = 0$ for identification. Assuming that the collateral variable and the item responses can be scored metrically, the log-multiplicative association model for the interaction of each of the item responses x_i with a collateral variable is

$$\lambda_{x_i z_j}^{X_i Z_j} = \mu_{x_i}^{X_i} \phi^{X_i Z_j} \nu_{z_j}^{Z_j}, \tag{9.16}$$

with constraints

$$\mu_0^{X_i} = 0 \quad \text{and} \quad \textstyle\sum_{z_j} \nu_{z_j}^{Z_j} = 0,$$

and

$$(\mu_1^{X_i})^2 = 1 \quad \text{and} \quad \textstyle\sum_{z_j} (\nu_{z_j}^{Z_j})^2 = 1.$$

In Equation 9.16, the parameter $\mu_{x_i}^{X_i}$ is a category score assigned to category x_i of measure i, $\nu_{z_j}^{Z_j}$ is a category score for category z_j of collateral variable j, and the parameter $\phi^{X_i Z_j}$ describes the degree of association between X_i and Z_j. If the restriction in Equation 9.16 is to be used in exchangeability models, we have to restrict all association parameters $\phi^{X_i Z_j}$ to be equal over items because, by symmetry, exchangeable scores should all have the same associations with the collateral variable(s). So $\phi^{X_i Z_j} = \phi^{X Z_j}$, where X denotes an arbitrary item. Furthermore it is assumed that the associations between item responses and the collateral variable are additive, that is, there are no higher order interactions between z_j and the item responses. Using Equation 9.16 under these restrictions we have the metric JCME model

$$\log P(\mathbf{x}z_j) = \lambda + \lambda_{\mathbf{x}} + \lambda_{z_j} + \sum_i \mu_{x_i}^{X_i} \phi^{XZ_j} \nu_{z_j}^{Z_j}$$
$$= \lambda + \lambda_{\mathbf{x}} + \lambda_{z_j} + t_1 \phi^{XZ_j} \nu_{z_j}^{Z_j},$$

where $t_1 = \sum_i u_i$ with $u_i = \mu_{x_i}^{X_i}$. Thus, in this model the $t_1 \times z_j$ interaction has been restricted to conform to a log-multiplicative association model for the item by collateral variable interactions. The theory for the estimation and testing of log-multiplicative association models has been described by Goodman (1986).

The second way to obtain nondegenerate metric exchangeability models is to restrict $\lambda_{\mathbf{t}}$ and $\lambda_{\mathbf{tz}}$ by assuming an underlying latent variable. It is easily shown from the JCME (Equation 9.8) or the SCME (Equation 9.9) LLM that

$$P(\mathbf{t}|\mathbf{z}) \propto \exp(\lambda_{\mathbf{t}} + \lambda_{\mathbf{tz}}).$$

As a result, restrictions on $P(\mathbf{t}|\mathbf{z})$ will generally lead to restrictions on $\lambda_{\mathbf{t}} + \lambda_{\mathbf{tz}}$. One way to restrict $P(\mathbf{t}|\mathbf{z})$ is to specify some distribution function for $P(\boldsymbol{\theta}|\mathbf{z})$ in Equation 9.15, for example, the multivariate normal distribution.

If one has sufficient confidence in the distribution of the latent variables, metric exchangeability models may be formulated for G_L. They would involve a latent variable and conditional independence assumptions of the item responses given the latent trait. To estimate latent variable LLMs, the distribution of θ is usually approximated by a discrete multinomial with probabilities and scale points fixed according to Gauss-Hermite Quadrature. By taking enough scale points θ can be approximated to any degree of precision by a log-linear latent-class model (Bock & Aitkin, 1981, Heinen, 1993). Heinen (1993) describes more general log-multiplicative latent variable models to which these restricted log-linear latent-class models apply. It is a subject of further research to assess to what log-linear latent-variable models are consistent with exchangeability models.

Almost all models in this chapter can be estimated and tested with the programmes ℓEM (Vermunt, 1997a, 1997b), LOGIMO (Kelderman, 1992, Kelderman & Steen, 1988), or DIGRAM (Kreiner, 1992, 2003). ℓEM is useful in the metric case, LOGIMO in the nominal cases. Graphical models, can be queried with the program DIGRAM. DIGRAM can also be used to generate ℓEM or LOGIMO setups for log-linear models that are consistent with certain graphical models.

9.7 Discussion

In this chapter we described measurement models that are suitable to test whether a set of items measure the same attribute. The development of CACs was based both on the relation between the item responses and their relations with variables in the nomological net. At various points it became apparent that *the degree to which a CAC has substantive validity depends on the extend to which the graphical model is representative for the nomological net around the attribute to be measured.* Therefore we would like to discourage the practice—usually unnecessary—to limit the application of a CAC to the relations between the item responses only. In that case, differential item functioning may go unnoticed and the result may be an invalid or unfair test.

It is particularly important to have a graphical model that is representative of the nomological net if normal distributions are assumed for the latent traits. Suppose that in the dementia example the variable $Z_2 = 1, 2$, dementia as diagnosed from neurological measurements, has a substantial influence on the item responses. If Z_2 was not in the graph, some latent trait distributions $P(\theta|\mathbf{z})$ would probably be bi-modal with peaks for the demented and the nondemented subpopulation. Thus, care should be exercised in making decisions about the type of distribution $P(\theta|\mathbf{z})$ if the graph does not sufficiently cover the nomological net. However, if the nomological net contains the most influential collateral variables, the normal distribution may be justified by the central limit theorem (Billingsley, 1995, chap. 6) if there are many omitted collateral variables that are approximately independent and identically distributed.

If a sufficiently large part of the nomological net is measured, theory driven researchers may abstain from modelling the relations between the item responses. In applied fields such as Education and Personnel Selection where fairness of individual scores is an issue, in addition to differential item functioning, systematic measurement errors due to test-taking artifacts are deemed more problematic. In that case, the CAC should also be sensitive to deviations of the expected relations between item responses.

References

Andersen, E. B. (1973). Conditional inference and multiple choice questionnaires. *British Journal of Mathematical and Statistical Psychology, 26*, 31-44.

Billingsley, P. (1995). *Probability and measure*. New York: Wiley.

Bock, R. D., & Aitkin, M. (1981). Marginal maximum likelihood estimation of item parameters: Application of an EM algorithm. *Psychometrika, 46*, 443-459.

Borsboom, D. (2003). *Conceptual issues in psychological measurement.* Unpublished doctoral dissertation, University of Amsterdam.

Cox, D. R., & Wermuth, N. (1996). *Multivariate dependencies–models, analysis and interpretation.* London: Chapman & Hall.

Cressie, N., & Holland, P. W. (1983). Characterizing the manifest probabilities of latent trait models. *Psychometrika, 48*, 129-142.

Cronbach, L. J. (1951). Coefficient alpha and the internal structure of tests. *Psychometrika, 16*, 297-334.

Cronbach, L. J., & Meehl, P. E. (1955). Construct validity in psychological tests. *Psychological Bulletin, 52*, 281-302.

DeFinetti, B. (1937). La prévision: Ses lois logiques, ses resources subjectives. [Prediction: Its logical laws, its subjective resources.] *Ann. Inst. Henri Poincaré, 7*, 1-68.

Duncan, O. D. (1984). Rasch measurement: Further examples and discussion. In C. F. Turner & E. Martin (Eds.), *Surveying subjective phenomena, Vol. 2* (pp. 367-403). New York: Russell Sage Foundation.

Fischer, G. H., & Molenaar, I. W. (Eds.). (1995). *Rasch models: Foundations, recent developments and applications.* New York: Springer.

Goodman, L. A. (1986). Some useful extensions of the usual correspondence analysis approach and the usual loglinear models approach in the analysis of contingency tables. *International Statistical Review, 54*, 243-309.

Gulliksen, H. (1968). Methods for determining equivalence of measures. *Psychological Bulletin, 70*, 534-544.

Haberman, S. J. (1979). *Analysis of qualitative data: Vol. 2, New developments.* New York: Academic Press.

Hambleton, R. K., & Van der Linden, W. J. (1997). *Handbook of modern item response theory.* New York: Springer.

Heinen, T. (1993). *Discrete latent variable models.* Tilburg, The Netherlands: Tilburg University Press.

Hempel, C. G. (1965). *Aspects of scientific explanation.* New York: The Free Press.

Hout, M., Duncan, O. D., & Sobel, M. E. (1987). Association and heterogeneity: structural models of similarities and differences. *Sociological Methodology, 17*, 145-184.

Huynh, H. (1978). Reliability of multiple classifications. *Psychometrika, 43*, 317-325.

Jensen, A. R. (1998). *The g-factor: The science of mental ability.* London: Westport.

Kelderman, H. (1984). Loglinear Rasch model tests. *Psychometrika, 54,* 223-245.

Kelderman, H. (1992). Computing maximum likelihood estimates of log-linear models from marginal sums with special attention to loglinear item response theory. *Psychometrika, 57,* 437-450.

Kelderman, H. (in press). Measurement exchangeability and normal one-factor models, *Biometrika.*

Kelderman, H., & Steen, R. (1993). LOGIMO - loglinear and loglinear IRT model analysis. [Computer program and manual.] Retrieved from *http://www.assess.com/Software/logimo.htm.*

Klein, J. P., Keiding, N., & Kreiner, S. (1995). Graphical models for panel studies, illustrated on data from the Framingham heart-study. *Statistics in Medicine, 14,* 1265-1290.

Kreiner, S. (1992). DIGRAM (Version 2.10). [Computer software and manual.] Retrieved from *http://www.biostat.mcw.edu/software/digram.html.*

Kreiner, S. (2003). *Introduction to DIGRAM.* University of Copenhagen, Department of Biostatistics, Report no. 10 Retrieved from *http://www.pubhealth.ku.dk/bsa/publ-e.htm.*

Lazarsfeld, P. F. (1959). Evidence and inference in social research. In D. Lerner (Ed.), *Evidence and inference* (pp. 107-138). Clencoe, IL: The Free Press.

Mokken, R. J. (1971). *A theory and procedure of scale analysis.* The Hague: Mouton/Berlin: De Gruyter.

Rasch, G. (1960). *Probabilistic models for some intelligence and achievement tests.* Copenhagen, Denmark: The Danish Institute for Educational Research. Expanded Edition (1980), Chicago: The University of Chicago.

Raven, J., Raven, J. C., & Court, J. H. (1991). *Manual for Raven's progressive matrices and vocabulary scales (section I): General overview.* Oxford, England: Oxford Psychologists Press.

Schuster, C. (2001). Kappa as a parameter of a symmetry model for rater agreement. *Journal of Educational and Behavioral Statistics, 26,* 331-342.

Spearman, C. (1910). Correlation calculated from faulty data. *British Journal of Psychology, 3,* 271-295.

Vermunt, J. K. (1997a). *ℓEM*: A general program for the analysis of categorical data. [Computer software and manual.] Retrieved from *http://www.kub.nl/mto/software2.html.*

Vermunt, J. K. (1997b). *Log-linear models for event histories.* Beverly
   Hills, CA: Sage.

Chapter 10

The Nedelsky Model
for Multiple-Choice Items

Timo M. Bechger, Gunter Maris,
Huub H.F.M. Verstralen, and Norman D. Verhelst
Cito National Institute for Educational Measurement

10.1 Introduction

Traditionally multiple choice (MC) items are scored binary. One point is earned when the correct answer is chosen and none are when any of the incorrect options (hereafter called "distractors") are chosen. This facilitates data analysis but it also entails the loss of information contained in the distracters (e.g., Levine & Drasgow, 1983). That is, we loose the information contained in the fact that there are different ways to answer an item incorrectly.

There have been various attempts to include the incorrect options into an item response theory (IRT) model (e.g., Bock, 1972; Thissen & Steinberg, 1984). Here, we discuss a new IRT model that we call *the Nedelsky model (NM)*. We focus on the psychometric properties of the NM and how they can be tested, and discuss a practical application. The reader is referred to research papers by Verstralen (1997a), and Bechger, Maris, Verstralen, and Verhelst (2003) for details about the estimation of the model by means of an EM-algorithm.

The NM derives its name from a method for standard setting suggested by Leo Nedelsky in 1954. Nedelsky's method is based on the idea that a person on the borderline between passing and failing responds to an MC question by first eliminating the answers he or she recognizes as wrong and then guesses at random from the remaining answers. The NM generalizes this idea in the sense that the selection of the answers is probabilistic and applies to all levels of ability. As in Nedelsky's method, it is assumed that the correct alternative is never rejected; that is, respondents will never think that the correct answer is wrong. Clearly, the NM is a restrictive model and we indicate how some of its assumptions may be relaxed.

Section 10.2 provides a brief description of the NM. Some of the psychometric properties of the model are discussed in detail in section 10.3 and we provide an informal procedure to evaluate them in section 10.4. In section 10.5, the NM is related to the two-parameter logistic (2PL) model and the three-parameter logistic (3PL) model. In section 10.6, the NM is applied to real data to see how it works in practice. Section 10.7 concludes the chapter. Mathematical proofs are in an Appendix to enhance the readability of the chapter

10.2   The Nedelsky Model

Consider an MC item i with $J_i + 1$ options arbitrarily indexed $0, 1, \ldots, J_i$. For convenience, 0 indexes the correct alternative. Let the random variable S_{ij} indicate whether alternative j is recognized to be wrong, and define $\mathbf{S}_i$ by the vector $(0, S_{i1}, \ldots, S_{iJ_i})$. We refer to $\mathbf{S}_i$ as *a latent subset*. The first entry in $\mathbf{S}_i$ is fixed at 0 because it is assumed that the correct alternative is never rejected. We comment on this assumption in the Discussion. The random variable $S_i^+ \equiv \sum_{j=1}^{J_i} S_{ij}$ denotes the number of distracters that are recognized as wrong.

The probability that alternative answer j is recognized as *wrong* by a respondent with ability θ is modelled as

$$\Pr(S_{ij} = 1|\theta) = \frac{\exp(a_i\theta - \zeta_{ij})}{1 + \exp(a_i\theta - \zeta_{ij})},$$

where ζ_{ij} represents the difficulty to recognize that option j of item i is wrong; $S_{i0} = 0$ implies that $\zeta_{i0} = \infty$. The discrimination parameter a_i is assumed to be positive so that $E[S_i^+|\theta] = \sum_{j=1}^{J_i} \Pr(S_{ij} = 1|\theta)$ is increasing in θ. It is helpful to think of each distracter as a dichotomous 2PL item where a correct answer is produced if the distracter is seen to be wrong.

As explained in the Introduction, the process that generates the response is assumed to consist of two stages. In the first stage, a respondent eliminates the distracters he recognizes to be wrong. Formally, this means that he draws a latent subset from the set of possible subsets $\Omega_{\mathbf{S}_i}$. Assuming independence among the options given θ, the probability that a subject with ability θ chooses any latent subset $\mathbf{s}_i \in \Omega_{\mathbf{S}_i}$ is given by the likelihood of J_i independent 2PL items; that is

$$
\begin{aligned}
\Pr(\mathbf{S}_i = \mathbf{s}_i|\theta) &= \prod_{j=1}^{J_i} \frac{\exp\left(a_i\theta - \zeta_{ij}\right)^{s_{ij}}}{1 + \exp\left(a_i\theta - \zeta_{ij}\right)} \\
&= \frac{\exp\left(a_i\theta s_i^+ - \sum_{j=1}^{J_i} s_{ij}\zeta_{ij}\right)}{\prod_{j=1}^{J_i}\left[1 + \exp\left(a_i\theta - \zeta_{ij}\right)\right]},
\end{aligned}
$$

where the sum $\sum_{j=1}^{J_i} s_{ij}\zeta_{ij}$ could be interpreted as a location parameter for the subset $\mathbf{s}_i$. If one suspects that there are dependencies among the options, a more appropriate model can be used instead (e.g., LOGIMO, Kelderman, 1984; The chain NM, Verstralen, 1997b).

Once a latent subset is chosen, a respondent is assumed to guess at random from the remaining answers. Thus, the conditional probability of responding with option j to item i, given latent subset $\mathbf{s}_i$, is given by

$$
\Pr(X_i = j|\mathbf{S}_i = \mathbf{s}_i) = \frac{1 - s_{ij}}{v(s_i^+)},
$$

where $X_i = j$ denotes the event that the respondent chooses alternative j, and $v(s_i^+) \equiv \sum_{h=0}^{J_i}(1 - s_{ih}) = J_i + 1 - s_i^+$ the number of alternatives to choose from. If a respondent knows the correct answer, $s_{ij} = 1$ for $j \neq 0$, $v(s_i^+) = 1$, and $\Pr(X_i = 0|\mathbf{S}_i = (0, 1, \ldots, 1)) = 1$. For later reference, $\Pr(X_i = j|\mathbf{S}_i = \mathbf{s}_i)$ is called *the response mapping*. Note that once a subset is chosen, each alternative in the subset is equally likely to be chosen. This assumption can be relaxed by changing the response mapping as in Equation 10.3, below. Note further that the second stage involves a randomization not involving θ, and hence can carry no information about θ.

Combining the two stages of the response process, we find that the conditional probability of choosing option j with item i is equal to

$$
\Pr(X_i = j|\theta) = \sum_{\mathbf{s}_i} \frac{1 - s_{ij}}{v(s_i^+)} \Pr(\mathbf{S}_i = \mathbf{s}_i|\theta).
$$

It is assumed that respondents independently choose subsets for different items so that the item responses are independent given θ. This is called *local stochastic independence* (LI).

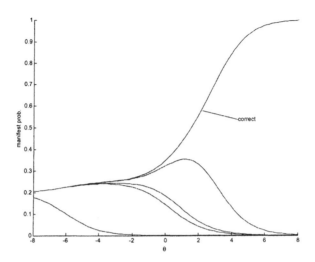

Figure 10.1: Conditional Probability of Choosing Option j of Item i as a Function of θ; $j = 0, \ldots, 4$.

There are four properties of the model that are readily seen in Figure 10.1. First, $\Pr(S_i^+ = 0|\theta) \to 1$ if $\theta \to -\infty$ which implies that

$$\lim_{\theta \to -\infty} \Pr(X_i = j|\theta) = \frac{1}{J_i + 1}, \text{ for } j = 0, \ldots, J_i. \qquad (10.1)$$

Second, if $\theta \to \infty$, $\Pr(S_i^+ = J_i|\theta) \to 1$ and

$$\lim_{\theta \to \infty} \Pr(X_i = 0|\theta) = 1.$$

Third,

$$\Pr(X_i = 0|\theta) - \Pr(X_i = j|\theta) = \sum_{\mathbf{s}_i} \frac{s_{ij}}{v(s_i^+)} \Pr(\mathbf{S}_i = \mathbf{s}_i|\theta) > 0$$

and the probability of a correct response is always larger than the probability to choose a distracter. Finally, Figure 10.1 suggests that $\Pr(X_i = 0|\theta)$ is an increasing function of θ and it can be proven that this is indeed the case (see Corollary 1).

For later reference, we note that because: (a) the latent trait is unidimensional, (b) the probability of a correct response is an increasing function of θ, and (c) LI is assumed, the NM pertains to the class of *monotone IRT models* (Mokken, 1971; Stout, 2002 and references therein).

10.3   Monotone Options Ratio

Option ratios are defined as

$$\psi_{itj}(\theta) \equiv \frac{\Pr(X_i = t|\theta)}{\Pr(X_i = j|\theta)},$$

where j and t index different answer alternatives. A model is said to have *monotone option ratios* (MOR) if all option ratios are monotone in θ. An equivalent definition is that

$$\frac{\Pr(\theta|X_i = t)}{\Pr(\theta|X_i = j)} \propto \varphi_{itj}(\theta)$$

is a monotone function of θ.

Theorem 1 *The Nedelsky model has MOR.*

The NM can be shown to have MOR. Specifically, it can be shown that $\psi_{itj}(\theta)$ is increasing in θ for any t and j such that $\zeta_{it} > \zeta_{ij}$. As mentioned in section 10.2, ζ_{i0} may formally be considered infinite so that $\psi_{i0j}(\theta)$ is always increasing in θ. If we consider the second definition of MOR this means that, if $\zeta_{it} \geq \zeta_{ij}$, $\theta|X_i = t$ is larger than $\theta|X_i = j$ in the sense of likelihood ratio (e.g., Ross, 1996, chap. 9).

We next mention two properties of the NM that are a consequence of MOR:

Corollary 1 $\Pr(X_i = 0|\theta)$ *is an increasing function of θ.*

Corollary 2 *If $\zeta_{it} \geq \zeta_{ij}$, $\Pr(X_i = t|\theta) \geq \Pr(X_i = j|\theta)$.*

Corollary 1 states that $\Pr(X_i = 0|\theta)$ is an increasing function of θ and this is true for any model which has MOR. Corollary 2 implies that $\Pr(X_i = t) > \Pr(X_i = j)$ if $\zeta_{it} > \zeta_{ij}$ so that the ordering among the option parameters can be inferred from the marginal probabilities. Recall that proofs are in the Appendix.

10.4   Graphical Investigation of Model Fit

Consider a test with I items. Let $X_+^{(-i)}$ denote the number of correct responses on the test excluding item i. The following lemma provides the key to investigate whether MOR is valid assuming only that the underlying model is a monotone IRT model.

Lemma 1 *Let $s_2 > s_1$. When the model is a monotone IRT model*

$$E[g(\theta)|X_+^{(-i)} = s_2] \geq E[g(\theta)|X_+^{(-i)} = s_1]$$

for all increasing functions g.

Let $X_i \in \{t, j\}$ denote the event that a respondent chooses either option t or option j. Note that

$$
\begin{aligned}
\Pr(X_i &= t|X_i \in \{t, j\}, X_+^{(-i)} = s). \\
&= \int \Pr(X_i = t|X_i \in \{t, j\}, \theta) f(\theta|X_+^{(-i)} = s) d\theta \\
&= E[\Pr(X_i = t|X_i \in \{t, j\}, \theta)|X_+^{(-i)} = s].
\end{aligned}
$$

MOR implies that

$$
\begin{aligned}
\Pr(X_i = t|X_i \in \{t, j\}, \theta) &= \frac{\Pr(X_i = t|\theta)}{\Pr(X_i = t|\theta) + \Pr(X_i = j|\theta)} \\
&= \frac{\psi_{itj}(\theta)}{1 + \psi_{itj}(\theta)}
\end{aligned}
$$

is a monotone function of θ. It follows from Lemma 1 that the probability $\Pr(X_i = t|X_i \in \{t, j\}, X_+^{(-i)})$ is monotone in $X_+^{(-i)}$.

In Table 10.1, $n(t, X_+^{(-i)})$ denotes the number of respondents that have chosen response t to item i with $X_+^{(-i)}$ correct responses to the other items. The proportion $\Pr(X_i = t|X_i \in \{t, j\}, X_+^{(-i)} = s)$ can be consistently estimated by the statistic

$$\frac{n(t, s)}{n(t, s) + n(j, s)}.$$

Plots of $X_+^{(-i)}$ against the estimates of $\Pr(X_i = t|X_i \in \{t, j\}, X_+^{(-i)})$ may be used to gain an impression of the validity of MOR. Note that there are $\frac{1}{2}(J_i + 1)J_i$ such plots for each item and one would usually look only at items and alternatives that are *a priori* considered suspicious to avoid capitalization on chance. Illustrations are given next, in section 10.5.

Table 10.1: Cross Tabulation of Total Score Deleting Item i $[X_+^{(-i)}]$ Against Item Score X_i.

			$X_+^{(-i)}$	
X_i	0	1	$\cdots$	I
0	$n(0,0)$	$n(0,1)$	$\cdots$	$n(0,I)$
$\vdots$	$\vdots$	$\vdots$	$\vdots$	$\vdots$
J_i	$n(J_i,0)$	$n(J_i,1)$	$\cdots$	$n(J_i,I)$

Note that many monotone IRT models might satisfy MOR without being anything like the NM. Thus, if MOR is not violated we should proceed to test the exact formulation of the NM before we may conclude that the NM fits the data. To this aim, we currently use plots of expected and observed probabilities for each of the response alternatives against estimated ability. Illustrations are given in section 10.5.

10.5   Relations to the 2PL and the 3PL

Consider a dichotomous item; that is, an item with two alternative answers one of which is correct answer. If the item is dichotomous, $J_i = 1$, and

$$\Pr(X_i = 0|\theta) = \tfrac{1}{2} + \left(1 - \tfrac{1}{2}\right)\Pr(S_{i1} = 1|\theta), \qquad (10.2)$$

where $\tfrac{1}{2} = \Pr(X_i = 0|S_{i1} = 0)$; the probability to find the correct answer by guessing, and $\Pr(S_{i1} = 1|\theta)$ is the probability that the respondent knows the correct answer. The probability to find the correct answer by guessing is fixed at $\tfrac{1}{2}$ because respondents who failed to eliminate the wrong answer find both alternatives equally attractive.

To relax this assumption, we could include parameters $\tau_{ij} > 0$ to represent the attractiveness of distracters relative to the correct alternative. In general, if $J_i \geq 1$, the response mapping would then become

$$\Pr(X_i = j|\mathbf{S}_i = \mathbf{s}_i; \tau_{i1}, \ldots, \tau_{iJ_i}) = \frac{(1 - s_{ij})\,\tau_{ij}}{\sum_{k=0}^{J_i}(1 - s_{ik})\tau_{ik}}, \qquad (10.3)$$

with $\tau_{i0} = 1$. Figure 10.2 illustrates the effect of the attractiveness parameters.

Consider again a dichotomous item. If Equation 10.3 is used as the

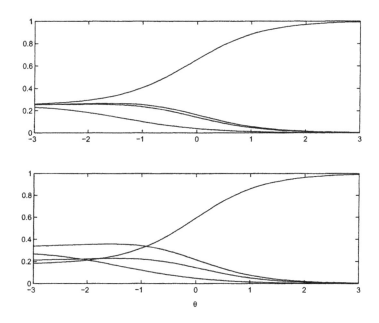

Figure 10.2: Conditional Probability of Choosing Option j of Item i as a Function of θ; $j = 0, \ldots, 4$. The Upper Figure Shows a NM. In the Lower Figure the Attractiveness Parameters of the Incorrect Alternatives are Greater than One. As a Consequence, Respondents With Low Abilities Have a Higher Probability Choosing an Incorrect Alternative Than Choosing the Correct Alternative.

response mapping, Equation 10.2 becomes

$$\Pr(X_i = 0|\theta) = \frac{1}{1 + \tau_{i1}} + \left(1 - \frac{1}{1 + \tau_{i1}}\right)\Pr(S_{i1} = 1|\theta),$$

which is equal to the 3PL. Thus, for dichotomous items, the NM is a special case of the 3PL where $\tau_{i1} = 1$, for all items. The 2PL is obtained when $\tau_{i1} \to \infty$; meaning that respondents who did not exclude the incorrect alternative will never choose the correct alternative by guessing. When there is more than one distracter ($J_i > 1$), the NM closely resembles the 3PL. That is, if we make no distinction between different incorrect responses, NM items can be fitted very closely by a 3PL.

10.6 An Application: The CITO Test 2003

To see how the NM works in practice we made use of data from the "CITO-test" which is administered at about 90% of Dutch elementary schools to children at the end of the final year of primary education. For our purpose we have taken a sample of 2000 children that took the test in the year 2003 and considered responses to a subset of 20 math items each with four alternative answers. An EM-algorithm was used for parameter estimation.

To assess the behavior of the items we prepared plots of expected and observed probabilities for each of the response alternatives against estimated ability. The lines represent expected probabilities although the symbols are observed probabilities in ability groups. The estimated abilities are posterior means.

The plots suggest that the NM is appropriate for the large majority of the items provided we allow differences in item discrimination. Typical plots are given in Figure 10.3 and Figure 10.4.

Among the items that fitted the NM well, there were quite a few where the distracters differ very little from one another (see Figure 10.4). Such items basically follow the 3PL. Items that do not appear to behave according to the NM show an interesting pattern that is seen in Figure 10.5.

In all these items, there was one distracter that was preferred over the correct alternative by children with relatively low ability respondents. With item 3, for instance less able children preferred $3\frac{1}{3}$ as an answer, and it is clear that they have been misled to think that $3\frac{1}{3}$ is the mean of $3\frac{1}{4}$ and $3\frac{1}{2}$ because 3 is the mean of 4 and 2. It would be more appropriate in this case to use a model with different attractiveness parameters as in Figure 10.2.

We plotted $X_+^{(-i)}$ against estimates of $\Pr(X_i = t|X_i \in \{t, j\}, X_+^{(-i)} = s)$ to investigate MOR. Values of $X_+^{(-i)}$ were grouped to ensure that there

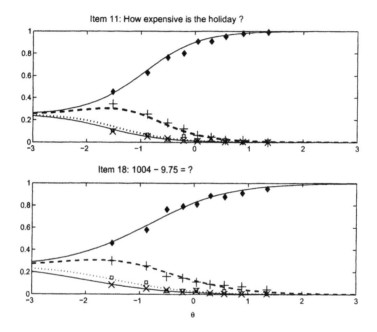

Figure 10.3: Plots of Observed Option Probabilities (Symbols) and Expected Option Probabilities (Curves) as a Function of Estimated Ability.

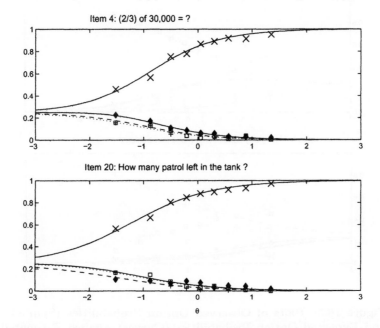

Figure 10.4: Plots of Observed Option Probabilities (Symbols) and Expected Option Probabilities (Curves) as a Function of Estimated Ability.

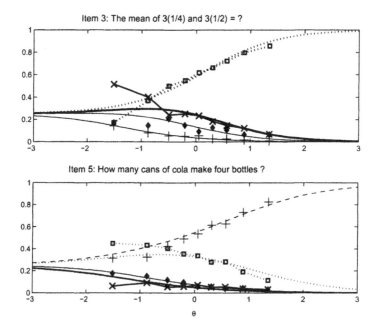

Figure 10.5: Plots of Observed Option Probabilities (Symbols) and Expected Option Probabilities (Curves) Against Estimated Ability. The Observed Option Probabilities are Connected and Can be Compared With the Curves Representing the Expected Option Probabilities. It May be Noted That for Some Options the Observed Probability May Be Higher or Lower Than Expected Under the NM.

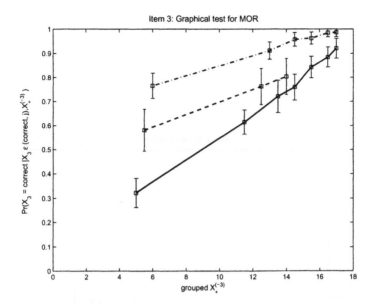

Figure 10.6: Plots of Total Score With Item 3 Deleted $[X_+^{(-3)}]$ Against an Estimate of $\Pr(X_3 = 0 | X_3 \in \{0, j\}, X_+^{(-3)})$, **for** $j = 1, 2, 3$.

would be at least 100 observations to calculate each proportion and approximate 95% confidence bounds were drawn based on a binomial distribution. Figure 10.6 and Figure 10.7 show these plots for item 3. No visible violations of MOR were found among the items.

Finally, one should note that most items were fairly easy and there was little data in the ability range where respondents take refuge to pure guessing. As a consequence, the present data were unsuited to test the NM against the 2PL which fitted the data well.

10.7 Conclusion

The NM is a restrictive model based on a simple theory about the response process. We have applied the NM to items from the CITO test and found that the model was adequate for most items. Violations of the NM were seen with items where respondents were misled to favor a wrong answer over the correct alternative. The introduction in the model of attractiveness

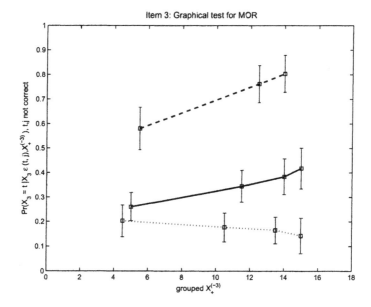

Figure 10.7: Plots of Total Score With Item 3 Deleted $[X_+^{(-3)}]$
Against an Estimate of $\Pr(X_3 = t|X_3 \in \{t,j\}, X_+^{(-3)})$ **for** $\{t,j\} =$
$\{1,2\}, \{1,3\}, \{2,3\}$.

parameters could serve to model such behavior, at the cost of increased sampling variability in all parameter estimates. Further extensions of the NM are discussed by Verstralen (1997b), and Bechger and Maris (2003). We have used informal graphical procedures to investigate the fit of the NM. Although we have found plots of expected and observed probabilities to be quite informative, it is clear that further tests must be developed before the NM could be brought in for our daily work.

We have assumed that the correct alternative is never rejected. Although this assumption may not appear very plausible, it can be shown that an absurd model results if we allow for (possibly empty) subsets without the correct alternative. Assume that the probability to include an alternative in the subset is increasing in θ for the correct alternative, and decreasing for any of the incorrect alternatives. Then, as θ decreases it becomes more likely that the subset will not contain the correct alternative. This means that a respondent with either a very high or a very low θ can determine which alternative is correct; it is either the only alternative in their subset (high ability) or the only alternative not in their subset (low ability). Stated otherwise, for some respondents it pays off to choose their response at random from the alternatives that are *not* in their subset; alternatives that they think are incorrect.

In the Introduction it was noted that binary scoring entails loss of information. Using the missing information principle (Orchard & Woodbury, 1972; Louis, 1982), it can be shown that binary scoring will indeed diminish the precision in estimated abilities to the extent that distracters differ in difficulty. It can, however, be shown that precision could be increased much more if we could somehow entice respondents to reveal their latent subsets. An illustration is given in Figure 10.8, which shows information functions for: (a) binary scoring, where it is registered whether the correct answer was chosen, (b) option scoring, where it is known what alternative was chosen, and (c) set scoring, where the latent subset was observed. It is seen that the information gain of option scoring relative to binary scoring is limited for the present application although set scoring is a more promising response format. Therefore, it is worthwhile to investigate whether set scoring could be applied in practice. Not only because it would increase information but also because it would, in principle, enable us to test the theory underlying the NM. It must be noted that set scoring has already been considered by others, although some time ago. For example, Coombs (1953) and Coombs, Milholland, and Wormer (1956) asked respondents to eliminate all options that they consider wrong. Alternatively, Dressel and Schmid (1953) asked respondents to indicate which options might be correct.

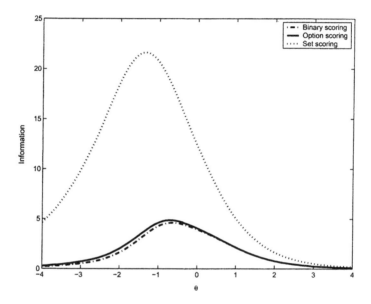

Figure 10.8: **Information Functions for Binary Scoring, Option Scoring, and Set Scoring Based on Parameter Values Obtained From the Illustration Discussed in Section 10.6.**

Appendix

To proof that the NM has MOR we use the following lemmas:

Lemma 1 *Assume that $J_i > 1$ and let $\Omega_{\mathbf{S}_i^{(j,t)}}$ denote the set of latent subsets without alternative j and t. Then*

$$\frac{\Pr(X_i = j|\theta)}{\Pr(S_{ij} = 0|\theta)} = \sum_{\mathbf{s}_i \in \Omega_{\mathbf{S}_i^{(t,j)}}} \left[\frac{\Pr(S_{it} = 1|\theta)}{v(s_i^+) + 1} + \frac{\Pr(S_{it} = 0|\theta)}{v(s_i^+) + 2}\right] \Pr(\mathbf{S}_i = \mathbf{s}_i|\theta).$$

Note that the ratio on the left side equals the probability that $X_i = j$ conditional on θ and given that the alternative is taken into consideration.

Proof: Let $\gamma_t(\theta; \mathbf{s}_i) \equiv \prod_{h \neq t}^{J_i} \Pr(S_{ih} = 0|\theta)^{1-s_{ih}} \Pr(S_{ih} = 1|\theta)^{s_{ih}}$, and $\gamma_{t,j}(\theta; \mathbf{s}_i) \equiv \prod_{h \neq t,j}^{J_i} \Pr(S_{ih} = 0|\theta)^{1-s_{ih}} \Pr(S_{ih} = 1|\theta)^{s_{ih}}$.

$$\Pr(X_i = j|\theta) = \sum_{\mathbf{s}_i} \frac{1 - s_{ij}}{v(s_i^+)} \prod_{h=1}^{J_i} \Pr(S_{ih} = 0|\theta)^{1-s_{ih}} \Pr(S_{ih} = 1|\theta)^{s_{ih}}$$

$$= \sum_{\mathbf{s}_i; s_{it}=1} (1 - s_{ij}) \frac{\Pr(S_{it} = 1|\theta)}{v(s_i^+)} \gamma_t(\theta; \mathbf{s}_i)$$

$$+ \sum_{\mathbf{s}_i; s_{it}=0} (1 - s_{ij}) \frac{\Pr(S_{it} = 0|\theta)}{v(s_i^+)} \gamma_t(\theta; \mathbf{s}_i)$$

$$= \sum_{\mathbf{s}_i \in \Omega_{\mathbf{S}_i^{(t)}}} (1 - s_{ij}) \frac{\Pr(S_{it} = 1|\theta)}{v(s_i^+)} \gamma_t(\theta; \mathbf{s}_i)$$

$$+ \sum_{\mathbf{s}_i \in \Omega_{\mathbf{S}_i^{(t)}}} (1 - s_{ij}) \frac{\Pr(S_{it} = 0|\theta)}{v(s_i^+) + 1} \gamma_t(\theta; \mathbf{s}_i)$$

$$= \sum_{\mathbf{s}_i \in \Omega_{\mathbf{S}_i^{(t)}}} (1 - s_{ij}) \left[\frac{\Pr(S_{it} = 1|\theta)}{v(s_i^+)} + \frac{\Pr(S_{it} = 0|\theta)}{v(s_i^+) + 1}\right] \gamma_t(\theta; \mathbf{s}_i)$$

$$= \sum_{\mathbf{s}_i \in \Omega_{\mathbf{S}_i^{(t)}}; s_{ij}=0} \Pr(S_{ij} = 0|\theta) \left[\frac{\Pr(S_{it} = 1|\theta)}{v(s_i^+)} + \frac{\Pr(S_{it} = 0|\theta)}{v(s_i^+) + 1}\right] \gamma_{t,j}(\theta; \mathbf{s}_i)$$

$$= \Pr(S_{ij} = 0|\theta) \sum_{\mathbf{s}_i \in \Omega_{\mathbf{S}_i^{(t,j)}}} \left[\frac{\Pr(S_{it} = 1|\theta)}{v(s_i^+) + 1} + \frac{\Pr(S_{it} = 0|\theta)}{v(s_i^+) + 2}\right] \Pr(\mathbf{S}_i = \mathbf{s}_i|\theta).$$

$\square$

Lemma 2

$$\frac{\partial}{\partial \theta}\psi_{itj}(\theta) = \frac{\Pr(X_i = t|\theta)\Pr(S_{ij} = 0|\theta) - \Pr(X_i = j|\theta)\Pr(S_{it} = 0|\theta)}{\Pr(X_i = j|\theta)^2}.$$

Proof: First,

$$
\begin{aligned}
\frac{\partial}{\partial \theta}\Pr(X_i = j|\theta) &= \sum_{\mathbf{s}_i}\frac{1 - s_{ij}}{v(s_i^+)}\frac{\partial}{\partial \theta}\Pr(\mathbf{S}_i = \mathbf{s}_i|\theta)\\
&= \sum_{\mathbf{s}_i}\frac{1 - s_{ij}}{v(s_i^+)}\Pr(\mathbf{S}_i = \mathbf{s}_i|\theta)\frac{\partial}{\partial \theta}\ln\Pr(\mathbf{S}_i = \mathbf{s}_i|\theta)\\
&= \sum_{\mathbf{s}_i}\frac{1 - s_{ij}}{v(s_i^+)}\Pr(\mathbf{S}_i = \mathbf{s}_i|\theta)\left(s_i^+ - E[S_i^+|\theta]\right)\\
&= \sum_{\mathbf{s}_i}\frac{1 - s_{ij}}{v(s_i^+)}\Pr(\mathbf{S}_i = \mathbf{s}_i|\theta)\left(E[v(S_i^+)|\theta] - v(s_i^+)\right)\\
&= E[v(S_i^+)|\theta]\Pr(X_i = 0|\theta) - 1 + \Pr(S_{ij} = 1|\theta).
\end{aligned}
$$

The fourth equality holds because $E[v(S_i^+)|\theta] - v(s_i^+) = J_i - E[S_i^+|\theta] + 1 - J_i + s_i^+ - 1 = s_i^+ - E[S_i^+|\theta]$. Now,

$$\frac{\partial}{\partial \theta}\psi_{itj}(\theta) = \frac{\frac{\partial}{\partial \theta}\Pr(X_i = t|\theta)\Pr(X_i = j|\theta) - \Pr(X_i = t|\theta)\frac{\partial}{\partial \theta}\Pr(X_i = j|\theta)}{\Pr(X_i = j|\theta)^2}$$

which simplifies to give the desired expression. $\qquad\Box$

Theorem 1 *The Nedelsky model has MOR.*

Proof: It follows from Lemma 2 that $\frac{\partial}{\partial \theta}\psi_{itj}(\theta)$ has the same sign as

$$\frac{\Pr(X_i = t|\theta)}{\Pr(S_{it} = 0|\theta)} - \frac{\Pr(X_i = j|\theta)}{\Pr(S_{ij} = 0|\theta)}. \tag{10.4}$$

Now, Lemma 1 implies that, for $J_i > 1$, Equation 10.4 equals

$$\sum_{\mathbf{s}_i \in \Omega_{\mathbf{s}_i^{(j,t)}}} c_{tj}(\theta, \mathbf{s}_i)\Pr(\mathbf{S}_i = \mathbf{s}_i|\theta),$$

where

$$
\begin{aligned}
c_{tj}(\theta, \mathbf{s}_i) &= \frac{\Pr(S_{ij} = 1|\theta)}{v(s_i^+) + 1} + \frac{\Pr(S_{ij} = 0|\theta)}{v(s_i^+) + 2} - \frac{\Pr(S_{it} = 1|\theta)}{v(s_i^+) + 1} - \frac{\Pr(S_{it} = 0|\theta)}{v(s_i^+) + 2}\\
&= (\Pr(S_{ij} = 1|\theta) - \Pr(S_{it} = 1|\theta))\left[\frac{1}{v(s_i^+) + 1} - \frac{1}{v(s_i^+) + 2}\right].
\end{aligned}
$$

If $\zeta_{it} > \zeta_{ij}$, $\Pr(S_{it} = 1|\theta) < \Pr(S_{ij} = 1|\theta)$, and it follows that $\frac{\partial}{\partial\theta}\psi_{itj}(\theta) > 0$ for all θ. Using Equation 10.2 the reader may verify that, if $J_i = 1$, $\psi_{i0j}(\theta) = \frac{1}{2}\Pr(S_{i1} = 1|\theta)$, which is increasing in θ. We have now proven that the NM has MOR. □

Corollary 1 $\Pr(X_i = 0|\theta)$ *is an increasing function of* θ.

Proof: MOR implies that

$$\sum_{j=1}^{J_i} \varphi_{ij0}(\theta) = \Pr(X_i = 0|\theta)^{-1}(1 - \Pr(X_i = 0|\theta))$$

is decreasing in θ. It follows that,

$$\frac{\partial}{\partial\theta}\Pr(X_i = 0|\theta)^{-1}(1 - \Pr(X_i = 0|\theta)) = -\frac{\frac{\partial}{\partial\theta}\Pr(X_i = 0|\theta)}{\Pr(X_i = 0|\theta)^2} < 0.$$

Hence, $\frac{\partial}{\partial\theta}\Pr(X_i = 0|\theta) > 0$ and $\Pr(X_i = 0|\theta)$ is increasing in θ. □

Corollary 2 *If* $\zeta_{it} \geq \zeta_{ij}$, $\Pr(X_i = t|\theta) \geq \Pr(X_i = j|\theta)$.

Proof: It follows from Equation 10.1 that $\lim_{\theta\to-\infty}\varphi_{itj}(\theta) = 1$. Hence, if $\zeta_{it} \geq \zeta_{ij}$ MOR implies that $\varphi_{itj}(\theta) \geq 1$ for all θ and the result follows. □

Lemma 3 *Let* $s_2 > s_1$. *When the model is a monotone IRT model,*

$$E[g(\theta)|X_+^{(-i)} = s_2] \geq E[g(\theta)|X_+^{(-i)} = s_1]$$

for all increasing functions g.

Proof: As we only register whether the answer was correct or not to calculate $X_+^{(-i)}$, it follows from Theorem 2 in Grayson (1988; see also Huynh, 1994) that $\theta|X_+^{(-i)} = s_2 \geq_{LR} \theta|X_+^{(-i)} = s_1$ if $s_2 > s_1$. Likelihood ratio ordering implies stochastic ordering which is equivalent to

$$E[g(\theta)|X_+^{(-i)} = s_2] \geq E[g(\theta)|X_+^{(-i)} = s_1]$$

for all increasing functions g (Ross, 1996, Proposition 9.1.2). □

References

Bechger, T. M., Maris, G., Verstralen, H. H. F. M., & Verhelst, N. D. (2003). *The Nedelsky model for multiple choice items.* R&D Report 03-05. Arnhem, The Netherlands: CITO.

Bechger, T. M., & Maris, G. (2003). *The componential Nedelsky model: A First Exploration.* R&D Report 03-03. Arnhem, The Netherlands: CITO.

Bock, R. D. (1972). Estimating item parameters and latent ability when responses are scored in two or more nominal categories. *Psychometrika, 37*-29-51.

Coombs, C. H. (1953). On the use of objective examinations. *Educational and psychological measurement, 13*, 308-310.

Coombs, C. H., Milholland, J. E., & Wormer, F. B. (1956). The assessment of partial knowledge. *Educational and Psychological measurement, 16*, 13-37.

Dressel, P. L., & Schmid, J. (1953). Some modifications of the multiple-choice test. *Educational and Psychological measurement, 13*, 574-595.

Grayson, D. A. (1988). Two-group classification in latent trait theory: Scores with monotone likelihood ratio. *Psychometrika, 53*, 383-392.

Huynh, H. (1994). A new proof of monotone likelihood ratio for the sum of independent Bernoulli random variables. *Psychometrika, 59*, 77-79.

Kelderman, H. (1984). Loglinear Rasch model tests. *Psychometrika, 49*, 223-245.

Levine, M. V., & Drasgow, F. (1983). The relation between incorrect option choice and estimated proficiency. *Educational and Psychological Measurement, 43*, 675-685.

Louis, T. A. (1982). Finding the observed information when using the EM algorithm. *Journal of the Royal Statistical Society, B, 44*, 226-233.

Mokken, R. J. (1971). *A theory and procedure of scale analysis.* The Hague: Mouton/Berlin: De Gruyter.

Nedelsky, L. (1954). Absolute grading standards for objective tests. *Educational and Psychological Measurement, 16*, 159-176.

Orchard, T., & Woodbury, M. A. (1972). A missing information principle: theory and applications. *Proceedings of the 6th Berkeley Symposium on Mathematical Statistics and Probability. 1*, 697-715.

Ross, S. M. (1996). *Stochastic processes* (2nd ed.). New York: Wiley.

Stout, W. F. (2002). Psychometrics: from practice to theory and back. *Psychometrika, 67*, 485-518.

Thissen, D., & Steinberg, L. (1984). A response model for multiple choice items. *Psychometrika, 49*, 501-519.

Verstralen, H. H. F. M. (1997a). *A latent IRT model for options of multiple choice items.* R&D Report 97-5. Arnhem, The Netherlands: CITO.

Verstralen, H. H. F. M. (1997b). *A logistic latent class model for multiple choice items.* R&D Report 97-1. Arnhem, The Netherlands: CITO.

Chapter 11

Application of the Polytomous Saltus Model to Stage-Like Proportional Reasoning Data

Karen Draney and Mark Wilson[1]
University of California, Berkeley

11.1 Introduction

Mixture IRT models are based on the assumption that the population be-ing measured is composed of two or more latent subpopulations, each of which responds to a set of tasks in predictably different ways. Within each subpopulation, a latent trait model holds for the entire set of tasks; however, between the subpopulations, there are differences that cannot be described within the constraints of the latent trait model used for a given subpopulation.

One of the most general mixture IRT models is the mixed Rasch model (Rost, 1990). This model assumes that the population in question is made up of N subpopulations, and that a Rasch model holds within each subpop-ulation. There is no necessary relation between the various Rasch models;

---
[1]We would like to thank Gerald Noelting for generously allowing us the use of his data.

the ordering of the items, for example, can be entirely different for each sub-population. This model is exploratory in the sense that it simply divides the population into the "best" (e.g., most different) set of subpopulations. The user must then determine what is interesting about the differences between subpopulations.

Other mixture IRT models are more confirmatory in nature (e.g., Mis-levy & Verhelst, 1990). Mislevy and Verhelst's model is an extension of the linear logistic test model (LLTM; Fischer, 1983). It posits a particular structure for item difficulty parameters within each subpopulation, based on characteristics of the tasks. A different LLTM may hold for each sub-population, if each responds differently to the task characteristics, or to a different set of task characteristics. This model is a special case of Rost's general model just described.

The saltus model (Wilson, 1984, 1989; Draney, 1996) is another such model. It was originally designed for the investigation of developmental stages. This model is a special case of both of the preceding models, with linear restrictions on the relationships between sets of item difficulties for the different subpopulations. It is generally assumed that the subpopula-tions are ordered in some way (as are developmental stages in children), and that groups of items become uniformly easier (or perhaps more diffi-cult) for higher subpopulations.

There are many educational and psychological theories that are based on the idea of developmental stages. Among these are some of the most in-fluential theories of human development of the twentieth century, including the theories of Jean Piaget (1950; Inhelder & Piaget, 1958), as well as more recent variations of this theory by such researchers as Siegler (1981), Bond (1995a; 1995b), Noelting (1980a; 1980b), Van Hiele (1986), and Demetriou and Efklides (1989).

For example, Siegler (1981) used the work of Piaget to show how sets of items associated with the skills acquired at different developmental stages changed in difficulty as children progressed through these stages. Some groups of items became easier and some more difficult, while others re-mained the same. The different developmental stages of the children thus resulted in relative shifts in the probability that certain groups of items would be answered correctly. The saltus model is suitable for use with such sets of items. A more general mixture IRT model, such as the mixed Rasch model, would require the estimation of a difficulty parameter for each item within each developmental stage (if the items are dichotomous); the saltus model can accommodate the developmental theory with one difficulty for each item, plus a small number of additional parameters to describe the changes associated with developmental stage.

A polytomous extension to the dichotomous model serves a number of purposes. First, it allows *partial credit* style scoring of developmental tasks (e.g., incorrect, partially correct, correct). Second, it allows one solution to a common potential cause of violation of local independence: replication of a certain task type with different specifics (e.g., the same problem structure but with different numbers, e.g., Rosenbaum, 1988). This solution involves summing the individual item scores to form a new, polytomous item (e.g., Wilson & Iventosch, 1988).

11.2   The Saltus Model

The saltus model is based on the assumption that there are H developmental stages in the population of interest. Items are constructed to represent each one of these stages, such that only persons at or above a stage are fully equipped to answer the items associated with that stage correctly. Once a person enters the developmental stage with which a set of items is associated, that person gains a substantial advantage in answering those items.

In the discussion that follows, the terms *person group* and *item class* will be used to differentiate between subpopulations of persons and sets of items designed to be suitable for such subpopulations. The saltus model assumes that all persons in group h answer all items in a manner consistent with membership in that group. However, persons within a group may differ by proficiency. In a Piagetian context, this means that a child in, say, the concrete operational stage is always in that stage, and answers all items accordingly. The child does not show formal operational development for some items and concrete operational development for others. However, some concrete operational children may be more proficient at answering items than are other concrete operational children.

To describe the model, suppose that, as in the partial credit model (Masters, 1982), the random variable X_{ni} indicating the response to item i (where i ranges from 1 to I) has $J_i + 1$ possible response alternatives or categories indexed $j = 0, 1, \ldots, Ji$. The parameter indicating step j in item i will be indicated by β_{ij}; the vector of all β_{ij} by $\boldsymbol{\beta}$.

Under the saltus model, an examinee is characterized not only by a proficiency parameter θ_n, but also by an indicator vector for group membership $\boldsymbol{\phi}_n$. If there are H potential person groups, $\boldsymbol{\phi}_n = (\phi_{n1}, \ldots, \phi_{nH})$, where ϕ_{nh} takes the value of 1 if the examinee n is in group h and 0 if not. The model assumes that each examinee belongs to one and only one group; thus, only one of the ϕ_{nh} is theoretically nonzero. As with θ_n, values of $\boldsymbol{\phi}_n$ are not observable.

Just as persons are associated with groups, and each person is a member of one and only one group, items are also associated with one and only one class. In a developmental context, an item's class would be said to be the first developmental stage at which a child would have all of the skills necessary to perform that item correctly. It is, of course, possible for children at lower developmental stages to perform the item correctly; however, this usually occurs because of guessing or a poorly developed strategy that happens to produce the correct answer in some cases. Unlike person group membership, however, which is unknown and must be estimated, item class membership is known a priori, based on the theory that was used to produce the items. It will be useful to denote item class membership by the indicator vector $\mathbf{b}_i$. As with θ_n, $\mathbf{b}_i = (b_{il}, \ldots, b_{iH})$, where b_{ik} takes the value of 1 if item i belongs to item class k, and 0 otherwise. The set of all $\mathbf{b}_i$ across all items is denoted by $\mathbf{b}$.

The equation

$$P(X_{nij} = j | \theta_n, \phi_{nh} = 1, \beta_i, \tau_{hk}) = \frac{\exp \sum_{s=0}^{j} (\theta_n - \beta_{is} + \tau_{hk})}{\sum_{t=0}^{J_i} \exp \sum_{s=0}^{t} (\theta_n - \beta_{is} + \tau_{hk})}$$

indicates the probability of response j to item i in a polytomous item response model that has been augmented by the introduction of the saltus parameter τ_{hk} as an additive element of the logistic argument. The saltus parameter describes the additive effect—positive or negative—for people in group h on the item parameters of all items in class k. Typically, in developmental contexts involving stages, this has taken the form of an increase in probability of success as the person achieves the stage at which an item is located, indicated by $\tau_{hk} > 0$ when $h \geq k$ (although this need not be the case). The saltus parameters can be represented together as an $H \times H$ matrix $\mathbf{T}$.

The probability that an examinee with group membership parameter ϕ_n and proficiency θ_n will respond in category j to item i is given by

$$P(X_{nij} = j | \theta_n, \phi_n, \beta_i, \mathbf{b}_i, \mathbf{T}) =$$
$$\prod_h \prod_k P(X_{nij} = j | \theta_n, \phi_{nh} = 1, \beta_i, \tau_{hk})^{\phi_{nh} b_{ik}}.$$

As item responses are assumed to be independent given θ_n, ϕ_n, and all of the item and saltus parameters, the modelled probability of a response vector is

$$P(\mathbf{X}_n = \mathbf{x}_n | \theta_n, \phi_n, \beta_i, \mathbf{b}_i, \mathbf{T}) =$$
$$\prod_h \prod_k \prod_i P(X_{nij} = x_{ij} | \theta_n, \phi_{nh} = 1, \beta_i, \tau_{hk})^{\phi_{nh} b_{ik}}.$$

The model requires a number of constraints on the parameters. For item step parameters, we choose to use two traditional constraints: first, $\beta_{i0} = 0$ for every item, and second, the sum of all the β_{ij} is set equal to zero. Some constraints are also necessary on the saltus parameters. This could be accomplished in several ways, but once parameters have been estimated with one set of restrictions, they can be translated to corresponding values under another set. The set of constraints we have chosen is the same as that used by Mislevy and Wilson (1996), and allow us to interpret the saltus parameters as changes relative to the first (lowest) developmental stage. Two sets of constraints are used. First, $\tau_{h1} = 0$; thus, the difficulty of the first class of items is held constant for all person groups; changes in the difficulty of classes of items greater than 1 are interpreted with respect to this first class of items for all person groups. Also $\tau_{1k} = 0$; thus, items as seen by person groups higher than 1 will be interpreted relative to the difficulty of the items as seen by person group 1.

Note that it would be possible to describe more general versions of this model in which the τ-parameters were indexed by item, or even by category within item (i.e., τ_{ihk} or τ_{ijhk}). However, many of the developmental theories that have been mentioned thus far are quite specific in terms of the effects they predict for developmental stage transitions on item performance. These strong theoretical predictions can often be captured in a smaller number of parameters. Rather than simply predicting that all items will change in difficulty when one enters another stage of development (in which case the mixed Rasch model might well be appropriate), theorists such as Piaget and Siegler predict that certain groups of tasks will change rather uniformly. The specific model described here, and used in the example to follow, was designed for just such predictions.

The EM algorithm (Dempster, Laird, & Rubin, 1977) was used for parameter estimation in the models discussed in this chapter.[1] This requires that some assumption be made about the distribution of proficiency within each subpopulation. For the analyses in this chapter, a normal distribution is assumed to hold within each person group. However, a semiparametric distribution (e.g., Mislevy & Wilson, 1996) could also be used.

Empirical Bayes estimation is used to produce estimates of the probability of group membership for each subject, as well as proficiency estimates given membership in each group. A person is said to be classified into the group for which his or her probability of membership is highest.

---

[1]Software used for parameter estimation was developed at UC Berkeley, and is currently available by contacting the first author directly at *k.draney@uclink4.berkeley.edu*

Figure 11.1: Example of Noelting Juice Mixtures Item. Note
That Dark Indicates Juice and Light Indicates Water.

11.3   An Example Application

The data for the first example are a set of responses to Noelting's (1980a;
1980b) Orange Juice Mixtures test for assessing proportional reasoning.
The items in this test consist of pictures of a certain number of glasses of
juice and glasses of water, representing a mixture. In each item, the child
is shown two such mixtures and asked which would taste more strongly of
juice, or if they would taste the same. A representation of such an item is
shown in Figure 11.1.

Noelting (1980a, 1980b) postulates a Piagetian stage hierarchy consisting of three stages, the intuitive, the concrete operational, and the formal
operational, for persons solving these items. Noelting develops juice mixture problems to represent the skills that differentiate between each developmental stage. In the intuitive stage, the child can additively compare the
relative quantity of an attribute (e.g., more glasses of juice or more glasses
of water), but tends to pay attention only to one attribute or the other. In
the concrete operational stage, the child begins to learn the concept of ratio
and proportionality. Rather than simply comparing the number of glasses
of juice or water between the two mixtures, the child is able to recognize
the concept of "one glass of juice for every glass of water" or "twice as
much juice as water." In the formal operational stage, the child learns to
deal formally with fractions, ratios, and percentages. Here, the child begins to master the formal mathematical rules for comparing two arbitrary
mixtures.

Noelting postulates three problem types (representing ordered substages)
within a stage and develops between one and four replications of each of
these substage problem types. These problem types and replications are
described in Table 11.1. The items were administered to a group of 460
subjects ranging in age from 5 to 17 years. The number of persons at each
age is given in Table 11.2.

Table 11.1: Noelting Problem Types.

S	IT	Description	Options
1	I	**Focusing on juice**. One mixture has more glasses of juice. Paying attention to juice alone will always produce the right answer.	4:1 vs. 1:4 2:1 vs. 1:2
	CO	**One-to-one equivalence**. One glass of juice for every glass of water in both mixtures.	1:1 vs. 2:2 2:2 vs. 3:3
	FO	**Multiplicity of denominators**. Proportionality relationship between juice and water can be set up in one mixture. Result of the same proportionality operation projected on the second pair.	3:1 vs. 5:2 8:3 vs. 3:1
2	I	**Focusing on water**. Glasses of juice equal. One mixture has more glasses of water.	1:0 vs. 1:1 1:2 vs. 1:3
	CO	**One-to-one equivalence**. First mixture has two glasses of juice for one of water (or vice versa). Second mixture is some multiple of this.	1:2 vs. 2:4 2:4 vs. 3:6
	FO	**Finding a common denominator**. Any two arbitrary mixtures of juice and water for which any previous strategy cannot easily be used.	5:2 vs. 7:3 3:5 vs. 5:8
3	I	**Compensation**. Less juice than water in one mixture; amounts are equal in the other mixture; or one mixture has more glasses of juice, other mixture has more glasses of water.	2:3 vs. 1:1 2:1 vs. 3:4
	CO	**Multiplication by** n. Ratio of juice to water in one mixture is n times that in the other.	4:3 vs. 8:6 3:1 vs. 6:2
	FO	**Addition of percents**. Involve adding mixtures of different strengths.	1:0:1 vs. 0:2:0 0:2:1 vs. 2:0:4 1:1:1 vs. 1:0:2 1:1:1 vs. 2:1:2

Note. S = Substage; It = Item type; I = Intuitive items; CO = Concrete operational items; FO = Formal operational items. If an item has options 2:3 vs. 1:1, then the respondent must choose between *two glasses of juice and three glasses of water* and *one glass of juice and one glass of water*. If an item has options 1:1:1 vs. 2:1:2, then the respondent must choose between *one glass of 40% juice, one glass of 10% juice and one glass of water* and *two glasses of 40% juice, one glass of 10% juice and two glasses of water*.

Table 11.2: Ages of Subjects in the Noelting Sample.

Age	Frequency	Percent
5	3	1
6	26	6
7	40	9
8	53	12
9	45	10
10	51	11
11	60	13
12	40	9
13	48	10
14	26	6
15	29	6
16	28	6
17	11	2
Total	460	

In cases such as this, where subsets of items are essentially replications of a single problem type, there is sometimes concern that the assumption of local independence of items will be violated. Although there are a variety of methods for dealing with such dependencies, one simple way to proceed is to sum the scores of the individual dichotomous items in a substage into a single polytomous item (Wilson & Iventosch, 1988). This results in three polytomous items representing each stage (and presumably ordered in difficulty level), with between two and five categories per item, for a total of nine polytomous items.

The saltus model to be fit to these data will be a three-stage model, comparing all three sets of items. In this model, saltus group 1 should include the youngest children in the intuitive stage, saltus group 2 should include middle-aged children in the concrete operational stage, and saltus group 3, the oldest children in the formal operational stage. In this model, there will be four between-group saltus parameters: One for concrete operational children taking concrete operational items, one for the same children taking the formal operational items, one for formal operational children taking concrete operational items, and one for formal operational children taking formal operational items. Two of these saltus parameters, those for concrete operational children taking concrete operational items and for formal operational children taking formal operational items, are expected to be positive. There are no specific predictions for the other two parameters.

Parameter estimates and standard errors for the this model are given in Table 11.3. Approximately two-thirds of the sample is classified into saltus group 1, and the majority of the remaining children (20% of the sample) is classified into group 2. Group 3 is small, comprising only 13% of the sample.

All saltus parameters are statistically different from zero (with magnitude more than twice their standard error), indicating that there is some systematic effect of class membership on item performance for both concrete and formal operational items. Recall that item difficulties as shown in Table 11.3 are interpreted relative to the lowest person group. The following is an example of how specific τ-parameters may be interpreted. For the lowest person group, intuitive item 1 has step parameters -3.60 and -7.68, whereas formal item 1 has step parameters 3.45 and 0.12. For person group 3, intuitive item 1 retains the same step parameters (although the mean proficiency of person group 3 is higher than for group 1, and thus the probability of correct responses to the intuitive items is higher for person group 3). However, the step parameters for formal item 1 are adjusted by τ_{33} when seen by person group 3 as follows: Step 1 becomes $3.45 - 2.10 = 2.35$, and step 2 becomes $0.12 - 2.10 = -1.98$. Not only are persons in group 3 more likely to score higher on all the items than are persons in group 1 (because of the higher average proficiency of group 3), the difference between the difficulties of intuitive and concrete items is greater for persons in group 1 than it is for persons in group 3. In probability terms, this means that an average person in group 1 has a .00 probability of scoring higher than 0 on formal item 1. If τ_{33} had been zero (and the difficulties of formal items had been the same for person group 3 as for person group 1), an average person in group 3 would have scored 1 with a probability of approximately .07, and 2 with a probability of .88. However, because of the size of τ_{33}, the average person scores 2 on this item with a probability of .99.

Average item difficulties for all nine items are shown in Table 11.4. Items are ordered in overall difficulty as predicted (average item difficulties for substage items increase within a stage; and items related to higher stages are on average more difficult than those for lower stages); however, there is little difference in the difficulty of concrete operational items two and three. This is not surprising, as the item type comprising concrete item 2 is actually just a subset of the item types comprising concrete item 3.

Groups are also ordered in mean proficiency as predicted. Group 1 has a low average proficiency estimate (-2.24 logits), group 2 has a moderately high average proficiency (1.79 logits), and group 3 the highest average proficiency (2.62 logits). The difference between groups 1 and 2 is more than 4 times greater than the difference between groups 2 and 3 (4.03 and 0.83

Table 11.3: Saltus Model Parameter Estimates and Standard Errors for the Noelting Model.

Parameter type	Parameter	Model notation	Estimate	SE
Intuitive item 1	Category 1	β_{11}	-3.60	(.854)
	Category 2	β_{12}	-7.86	(.712)
Intuitive item 2	Category 1	β_{21}	-4.09	(.294)
	Category 2	β_{22}	-4.50	(.294)
Intuitive item 3	Category 1	β_{31}	-2.43	(.218)
	Category 2	β_{32}	-3.96	(.188)
Concrete item 1	Category 1	β_{41}	$-.97$	(.178)
	Category 2	β_{42}	-2.25	(.193)
Concrete item 2	Category 1	β_{51}	.71	(.184)
Concrete item 3	Category 1	β_{61}	.83	(.241)
	Category 2	β_{62}	$-.59$	(.333)
	Category 3	β_{63}	2.36	(.205)
Formal item 1	Category 1	β_{71}	3.45	(.393)
	Category 2	β_{72}	.12	(.384)
Formal item 2	Category 1	β_{81}	3.85	(.301)
	Category 2	β_{82}	1.68	(.294)
Formal item 3	Category 1	β_{91}	7.15	(.342)
	Category 2	β_{92}	4.57	(.353)
	Category 3	β_{93}	1.84	(.352)
	Category 4	β_{94}	3.69	(—)
Saltus parameter	Group 2 class 2	τ_{22}	2.59	(.004)
	Group 2 class 3	τ_{23}	.93	(.003)
	Group 3 class 2	τ_{32}	.58	(.004)
	Group 3 class 3	τ_{33}	2.10	(.005)
Saltus group 1	Proportion	π_1	.66	(.019)
	Mean	μ_1	-2.24	(.062)
	Standard deviation	$\sigma 1$	1.16	(.042)
Saltus group 2	Proportion	π_2	.21	(.022)
	Mean	μ_2	1.79	(.104)
	Standard deviation	σ_2	1.06	(.057)
Saltus group 3	Proportion	π_3	.13	(—)
	Mean	μ_3	2.62	(.127)
	Standard deviation	σ_3	.95	(.084)

Table 11.4: Average Item Difficulties for the Noelting Model.

Item	Average difficulty
Intuitive 1	−5.73
Intuitive 2	−4.29
Intuitive 3	−3.20
Concrete op 1	−1.61
Concrete op 2	0.71
Concrete op 3	0.87
Formal op 1	1.79
Formal op 2	2.76
Formal op 3	4.31

logits, respectively).

As predicted, there is a positive saltus effect for persons in group 2 taking concrete operational items (2.59 logits with respect to group 1). There is a similar-size positive saltus effect for persons in group 3 taking formal operational items (2.10 logits with respect to group 1). In addition, there are smaller positive effects for persons in group 2 taking formal operational items (about one logit with respect to persons in group 1), and persons in group 3 taking concrete operational items (about half a logit with respect to persons in group 1).

The interpretation of item difficulties and mean abilities for groups is often easier when these parameters are displayed in a graphical form sometimes referred to as a Wright Map (see Wilson, in press) in honor of its creator: Benjamin D. Wright of the University of Chicago. Maps have long been used with Rasch-family models, such as the partial credit and rating scale models, and are incorporated into many estimation software packages for these models. A Wright Map of the mean group abilities and the item difficulties as seen by each group is given in Figure 11.2. In this figure, the units of the logit scale (the scale in which parameters for this model are estimated), are shown on the far left side of the figure. The column to the right of this contains the mean abilities of the person groups, with a range of one standard deviation on either side of each group mean. The mean of each group is represented by the letter M followed by the group number (e.g., M1 for the mean of group 1). Similarly, the upper and lower limits of the standard deviation range are represented by the letter S and the group number; these limits are connected by dashed lines to the group mean.

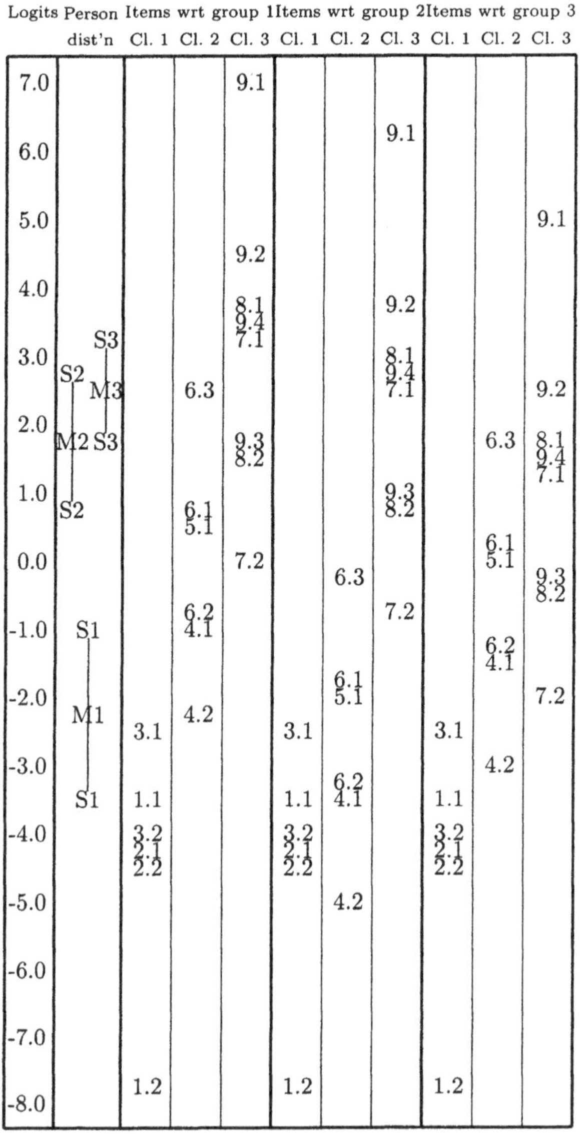

Logits	Person dist'n	Items wrt group 1			Items wrt group 2			Items wrt group 3		
		Cl. 1	Cl. 2	Cl. 3	Cl. 1	Cl. 2	Cl. 3	Cl. 1	Cl. 2	Cl. 3
7.0				9.1						
6.0							9.1			
5.0				9.2						9.1
4.0				8.1 / 9.4 / 7.1			9.2			
3.0	S3 / S2 / M3		6.3				8.1 / 9.4 / 7.1			9.2
2.0	M2 S3			9.3 / 8.2					6.3	8.1 / 9.4 / 7.1
1.0	S2		6.1 / 5.1				9.3 / 8.2			
0.0				7.2		6.3			6.1 / 5.1	9.3 / 8.2
-1.0	S1		6.2 / 4.1				7.2		6.2 / 4.1	
-2.0	M1	3.1	4.2		3.1	6.1 / 5.1		3.1		7.2
-3.0	S1	1.1			1.1	6.2 / 4.1		1.1	4.2	
-4.0		3.2 / 3.1 / 2.2			3.2 / 3.1 / 2.2			3.2 / 3.1 / 2.2		
-5.0						4.2				
-6.0										
-7.0										
-8.0		1.2			1.2			1.2		

Figure 11.2: Map of Noelting Three-Level Analysis.

The difficulty levels for the various item steps as seen by each group are shown in the remaining columns of Figure 11.2. For each group, the difficulties for each item class are separated by dashed line. The difficulty levels for the items as seen by group 1 are shown in the columns labelled "Items wrt to group 1" and similarly for groups 2 and 3. Under each of these headings, the items are broken into columns based on their class; intuitive items (items 1-3) are in the leftmost column under each heading; concrete items (items 4-6) are in the center column, and formal items (items 7-9) are in the rightmost column. Individual items steps are represented by the item number, a dot, and then the step number (e.g., 3.2 represents item 3, step 2). More difficult item steps and more able persons are toward the top of the figure, and less difficult item steps and less able persons are toward the bottom of the figure. The effect of the saltus parameters for all three groups can be seen in this figure: Although the difficulty of the intuitive items is held fixed, the difficulty of the formal items drops steadily across all three groups, and the difficulty of the concrete items first drops, then rises again somewhat. Both groups 2 and 3 have ability considerably above that of group 1, whereas the difference between them is somewhat smaller.

Modelled probabilities of response by a person whose proficiency was equal to the mean of each group, using the estimated parameter values, are given in Table 11.5. For the intuitive items, all three groups are most likely to score 2, and the probability of a score of 2 rounds to 1.00 on all three items for the two higher groups. Response probabilities for the concrete items are similar for the two higher groups, with the highest response the most probable in all cases. This is clearly different from group 1, for which the lowest response is most probable. The items that most clearly differentiate between groups 2 and 3 are the formal items. For group 1, the probability of response 0 rounds to 1.00 in all cases. Group 2 has some probability of higher responses to the formal items, but for all but the easiest of these items, response 0 still has the highest probability. For group 3, however, the most probable response to all items is the highest possible response, although for all but the intuitive and the first of the concrete items, there is a nonzero probability of some other response as well.

An example set of person response vectors, classification probabilities, ability estimates, and standard errors is listed in Table 11.6. Persons such as A or B, who scored any number of points on the intuitive items and no points on any other items, are clearly classified into group 1. This is also true for persons such as C or D, who scored one or two points on any of the concrete operational items (other than the most difficult one) in addition to the above points. If a person, in addition to answering most or all of the intuitive items correctly, also answered one or two of the subitems of concrete item 3 correctly, as did person E, that person began to have some

Table 11.5: Response Probabilities by Saltus Group for the Noelting Model.

Groups	Item	Item score				
		0	1	2	3	4
Group 1	Intuitive 1	.00	.00	1.00		
	Intuitive 2	.01	.09	.89		
	Intuitive 3	.11	.13	.75		
	Concrete 1	.64	.18	.18		
	Concrete 2	.95	.05			
	Concrete 3	.95	.04	.01	.00	
	Formal 1	1.00	.00	.00		
	Formal 2	1.00	.00	.00		
	Formal 3	1.00	.00	.00	.00	.00
Group 2	Intuitive 1	.00	.00	1.00		
	Intuitive 2	.00	.00	1.00		
	Intuitive 3	.00	.00	1.00		
	Concrete 1	.00	.00	1.00		
	Concrete 2	.02	.98			
	Concrete 3	.00	.00	.12	.88	
	Formal 1	.13	.06	.81		
	Formal 2	.45	.14	.41		
	Formal 3	.98	.01	.00	.00	.00
Group 3	Intuitive 1	.00	.00	1.00		
	Intuitive 2	.00	.00	1.00		
	Intuitive 3	.00	.00	1.00		
	Concrete 1	.00	.00	1.00		
	Concrete 2	.08	.92			
	Concrete 3	.00	.01	.30	.69	
	Formal 1	.00	.01	.99		
	Formal 2	.02	.04	.94		
	Formal 3	.12	.01	.01	.22	.63

probability of being classified into group 2. Persons who scored most or all of the possible points on both the intuitive and the concrete operational items, as with persons F and G, were classified solidly into group 2, although if one or two of the subitems of the easier concrete operational items were missed (as with person F), there was still some probability of belonging to group 1. Person H, who scored, in addition to most of the intuitive and concrete items, a few points on the formal operational items, was still most likely to be in group 2. With person I, a perfect score on all but formal item 3 yielded only a 17% probability of being in group 3. Even persons J and K, who had a perfect score or only missed one of the subitems of formal item 3, did not have over a 90% probability of being in group 3—the probabilities of being in group 2 were still 19% and 35%, respectively, although the probability of being in group 3 is now more than 60%. The only persons with a probability greater than 90% of being in group 3 were like person L, who answered the intuitive and most or all of the formal items correctly, and who missed points on some of the concrete items. Such persons, it seems, were classified into group 3 only because they did not fit well in either of the other two groups. In all cases, to be in one of the higher ability groups with probability greater than .5, it was necessary to have mastered most or all of the items representing that group.

Even unusual response vectors have an interesting classification story. For example, person M, who answered all of the intuitive items correctly, only the easiest of the concrete operational items, and all of the subitems of item 2 of the formal operational items, has a small probability of being in group 3, and no probability of being in group 2, although this person is still most likely a member of group 1. Such a response vector seems to indicate some degree of misfit of this person to the model, and it would be interesting to speak to the person, to see if the two correct answer to formal items were the result of lucky guesses or of an idiosyncratic strategy that happened to produce the correct result. In this data set, less than 1% of all cases showed "misfitting" patterns such as this, suggesting that the developmental theory tends to predict student performance well in nearly all cases.

Also of interest is the table of person classification by age, shown in Table 11.7. According to the original Piagetian theory, a person should enter the formal operational stage sometime during early adolescence, and should be solidly into this stage by their late teens. Persons in the middle and late formal operational stage (i.e., the persons in late adolescence and older), should be able to correctly solve, barring trivial errors due to carelessness, most or all of the items targeted to this stage: the problems involving addition of percents.

However, an examination of Table 11.7 shows clear differences between

Table 11.6: Example of Probabilities of Response Patterns for 13 Persons for the Noelting Model.

Person	Responses	Score	Probability	Ability	SE
A	000000000	0	1.00	−4.88	.60
		1	.00	−3.85	.52
		2	.00	−1.79	.24
B	222000000	0	1.00	−2.30	.66
		1	.00	−1.94	.62
		2	.00	−.78	.52
C	222100000	0	1.00	−1.86	.67
		1	.00	−1.52	.66
		2	.00	−.51	.52
D	222110000	0	1.00	−1.40	.67
		1	.00	−1.06	.68
		2	.00	−.23	.52
E	222202000	0	.89	−.51	.66
		1	.10	−.09	.68
		2	.01	.32	.53
F	222203000	0	.29	−.08	.64
		1	.69	.35	.66
		2	.01	.61	.53
G	222213000	0	.02	.32	.63
		1	.98	.78	.64
		2	.00	.90	.54
H	222213210	0	.00	1.31	.48
		1	.92	1.97	.61
		2	.08	1.73	.50
I	222213220	0	.00	1.51	.40
		1	.83	2.34	.59
		2	.17	1.99	.50
J	222213224	0	.00	1.85	.19
		1	.19	3.58	.55
		2	.81	3.33	.72
K	222213223	0	.00	1.80	.23
		1	.35	3.28	.54
		2	.65	2.87	.62
L	222201224	0	.00	1.65	.33
		1	.00	2.68	.56
		2	1.00	2.24	.51
M	222200020	0	.81	−.51	.66
		1	.00	−.09	.68
		2	.19	.32	.53

Table 11.7: Tabulation of Saltus Class by Age for the Noelting Model.

Age	Group 1	Group 2	Group 3
5	3	0	0
6	26	0	0
7	40	0	0
8	53	0	0
9	43	2	0
10	44	7	0
11	46	14	0
12	25	15	0
13	14	24	10
14	3	11	12
15	4	10	15
16	1	11	16
17	0	4	7
Totals	302	98	60

what one might predict based on a strict interpretation of Piaget's original theory, and what one observes in this data set. The table seems to indicate that the transition between the intuitive and concrete stages (as estimated by the model) happens at approximately age 12. This is the age at which persons begin to get most or all of the concrete operational items correct. Although examination of person classification does show that persons need to answer all or nearly all of the concrete operational items correctly to be classified into the concrete operational class, this is still rather surprising. The transition between the intuitive and the concrete operational stages should, according to Piaget and some neo-Piagetian researchers, happen much earlier than age 12, and it seems likely that 10 and 11-year-old youngsters should be able to answer most or all concrete operational items correctly. However, this does not happen in this data set. There were 142 persons age 13 and older. Whereas 90 (63.4%) of these persons answered all of the concrete items correctly, only 32 (22.5%) answered all of the formal items correctly. Of the oldest persons, those age 16 and 17, only 15 of 39 (38.5%), or substantially less than half, answered all of the formal items correctly, and 17 (43.6%) missed two or more points on these items.

This seems to indicate that either some of the persons at the older age levels in these data sets are not taking the tasks very seriously and are being rather careless, or that the difficulty of the tasks is higher than would be predicted by a strict interpretation of Piagetian theory. Some researchers believe that entry into the formal operational stage, and in particular the late formal operational stage, occur at older ages than Piaget's theory originally stated (cf. Bond, 1995a). Some theorists claim that not all adults reach the later substages of the formal operational stage (Bond, 1995b). The information in this data set lends at least some support to the contention that entrance into the formal operational developmental stage, and particularly the later substages of this stage, might happen later than strict interpretation of Piagetian theory would indicate.

A test of fit for the model against a general multinomial alternative yields a chi-square statistic of 4129.54, with degrees of freedom equal to the number of observed response patterns, minus the number of parameters estimated, minus one, which in this case is $65 - 31 - 1 = 33$. This is significant at the .01 level, indicating that the fitted saltus model results in significantly worse fit than an unrestrictive multinomial model on response counts.

11.4   Discussion

The use of the saltus model has allowed us to learn some interesting things about the example data set. For instance, it would seem that the saltus model is more suitable for use with this data than a latent class model. Latent class models are similar to mixture IRT models, in that they assume the observed population is composed of latent subpopulations; however, such models include no person parameters; group membership accounts for all explained variation between persons, and within-group variation is considered error. However, as seen in Table 11.4, each person group has a standard deviation of approximately one logit, indicating that there is some amount of within-group variability in these data. The saltus model is also similar to multilevel item response models, which include a latent regression model (see, e.g., Adams, Wilson, & Wu, 1997), using known person classification variables. One might, for example have classified persons into developmental stages based on age, and estimated effects on the various groups of items as regression coefficients. However, as Table 11.7 shows, this would likely lead to some inaccuracy. Although all children age 8 and under, and most who are 9 are classified into the lowest developmental stage, there is no other age at which all children are classified the same, and some ages (e.g., age 13) where there seem to be children at all three developmental stages. Testing against the unrestricted multinomial logit model indicated that the fit of the saltus model was significantly worse than that obtained by simply using a general multinomial model for response counts. Tests against an unrestricted model are not often terribly useful; they almost always indicate significant misfit, but provide little additional information. Tests of fit against other models, such as the partial credit model, would be informative; however, likelihood ratio tests between one-class IRT models and mixture IRT models are not valid due to boundary problems (Böhning, 1999). Further analysis of fit would prove useful. For example, it might be useful to develop a saltus-like model with variable item slopes, as models with equal slopes for all items are often too restrictive to fit well. In addition, it might be the case that models which included saltus parameters indexed by individual item and/or step, rather than simply associating saltus parameters with items as a whole, and estimating a single parameter across all items within an item group, might yield interesting differences by item and/or step.

References

Adams, R. J., Wilson, M., & Wu, M. (1997). Multilevel item response models: an approach to errors in variables regression. *Journal of Educational and Behavioral Statistics, 22,* 47-76.

Böhning, D. (1999). *Computer-Assisted analysis of mixtures and applications: Meta-analysis, disease mapping and others.* London: Chapman & Hall/CRC.

Bond, T. G. (1995a). Piaget and Measurement I: The twain really do meet. *Archives de Psychologie, 63,* 71-87.

Bond, T. G. (1995b). Piaget and Measurement II: Empirical validation of the Piagetian model. *Archives de Psychologie, 63,* 155-185.

Demetriou, A., & Efklides, A. (1989). The person's conception of the structures of developing intellect: Early adolescence to middle age. *Genetic, Social, and General Psychology Monographs, 115,* 371-423.

Dempster, A. P., Laird, N. M., & Rubin, D. B. (1977). Maximum likelihood from incomplete data via the EM algorithm. *Journal of the Royal Statistical Society, Series B, 39,* 1-38.

Draney, K. (1996). *The polytomous saltus model: A mixture model approach to the diagnosis of developmental differences.* Unpublished doctoral dissertation, University of California, Berkeley.

Fischer, G. H. (1983). Logistic latent trait models with linear constraints. *Psychometrika, 48,* 3-26.

Inhelder, B. & Piaget, J. (1958). *The growth of logical thinking from childhood to adolescence.* New York: Basic.

Masters, G. N. (1982). A Rasch model for partial credit scoring. *Psychometrika, 47*, 149-174.

Mislevy, R. J., & Verhelst, N. (1990). Modeling item responses when different subjects employ different solution strategies. *Psychometrika, 55*, 195-215.

Mislevy, R. J., & Wilson, M. (1996). Marginal maximum likelihood estimation for a psychometric model of discontinuous development. *Psychometrika, 61*, 41-71

Noelting, G. (1980a). The development of proportional reasoning and the ratio concept—Part I: Differentiation of stages. *Educational Studies in Mathematics, 11*, 217-253.

Noelting, G. (1980b). The development of proportional reasoning and the ratio concept—Part II: Problem-structure at successive stages; problem-solving strategies and the mechanism of adaptive restructuring. *Educational Studies in Mathematics, 11*, 331-363.

Piaget, J. (1950). *The psychology of intelligence.* (M. Piercy, Trans.) London: Lowe & Brydone. (Original work published 1947)

Rosenbaum, P. R. (1988). Item bundles. *Psychometrika, 53*, 349-359.

Rost, J. (1990). Rasch models in latent class analysis: An integration of two approaches to item analysis. *Applied Psychological Measurement, 14*, 271-282.

Siegler, R. S. (1981). Developmental sequences within and between concepts. *Monograph of the Society for Research in Child Development, 46*(1, Serial No. 189).

Van Hiele, P. M. (1986). *Structure and insight: A theory of mathematics education.* Orlando, FL: Academic Press.

Wilson, M. (1984). *A psychometric model of hierarchical development.* Unpublished doctoral dissertation, University of Chicago, Chicago.

Wilson, M. (1989). Saltus: A psychometric model of discontinuity in cognitive development. *Psychological Bulletin, 105*(2), 276-289.

Wilson, M. (in press). *Measurement: A constructive approach.* Mahwah, NJ: Erlbaum.

Wilson, M., & Iventosch, L. (1988). Using the partial credit model to investigate responses to structured subtests. *Applied Measurement in Education, 1*, 319-334.

Chapter 12

Multilevel IRT Model Assessment

Jean-Paul Fox
University of Twente

12.1 Introduction

Modelling complex cognitive and psychological outcomes in, for example, educational assessment led to the development of generalized item response theory (IRT) models. A class of models was developed to solve practical and challenging educational problems by generalizing the basic IRT models. An IRT model can be used to define a relation between observed categorical responses and an underlying latent trait, such as, ability or attitude. Subsequently, the latent trait variable can be seen as the outcome in a regression analysis. That is, a regression model defines the relation between the latent trait and the set of predictors. The combination of both models, a regression model imposed on the ability parameter in an IRT model, can be viewed as an extension to the class of IRT models.

Verhelst and Eggen (1989), and Zwinderman (1991, 1997) considered the combination of an IRT model with a structural linear regression model. Zwinderman showed that the correlation between the latent trait and other variables can be estimated directly without estimating the subject parameters. A straightforward extension of this model consists of a structural multilevel model imposed on the latent trait variable. Adams, Wilson and

Wu (1997) and Raudenbush and Sampson (1999) discussed a multilevel model that can be seen as a Rasch model embedded within a hierarchical structure, where the first level of the multilevel model describes the relation between the observed item scores and the ability parameters. This multilevel model can be estimated in HLM 5 (Raudenbush, Bryk, Cheong, & Congdon, 2000). A multilevel formulation of the Rasch model that can be estimated using the HLM software was developed by Kamata (2001). Maier (2001), defined a hierarchical Rasch model, that is, the person parameters in the Rasch model are modelled with a one-way ANOVA with random effects. Fox (in press) and Fox and Glas (2001) extended the two-parameter normal ogive model and the graded response model by imposing a multilevel model on the ability parameters with covariates on both levels. This multilevel IRT model describes the link between dichotomous or polychotomous response data and a latent dependent variable as the outcome in a structural multilevel model. This extension allows to model relationships between observed and latent variables on different levels using dichotomous and polytomous IRT models that relate the test performances to the latent variables. That is, relationships between abilities of students underlying the test and other observed variables of some individual or group characteristics can be analyzed taking into account the errors of measurement using dichotomous or polytomous indicators.

Verhelst and Eggen (1989) proclaimed a strict distinction between the estimation of the parameters of the measurement model and the structural model. One should first calibrate the measurement model before estimating the structural model parameters. This way it is possible to distinguish possible model violations in the measurement model and the structural model. Alternatively, a two-stage estimation procedure can cause biased parameter estimates and underestimation of some standard errors due to the fact that some parameters are held fixed at values estimated from the data, depending on the available calibration data. Furthermore, in educational testing the response data often have a hierarchical structure and the measurement model ignores this effect of the clustering of the respondents. In Fox (in press) and Fox and Glas (2001) a procedure was developed for estimating simultaneously all model parameters. This Bayesian method (Markov chain Monte Carlo, MCMC) handles all sources of uncertainty in the estimation of the model parameters.

The goal of this chapter is to develop methods to assess the plausibility of the model or some of the assumptions under the preferred Bayesian estimation method. The MCMC estimation procedure is time-consuming and it is, therefore, preferable to compute certain fit statistics during the estimation of the parameters or based on the MCMC output. In this chapter, methods are proposed for checking the fit of multilevel IRT models, which are simply byproducts of the MCMC procedure for estimating the model parameters. Samples of the posterior distributions are obtained from the MCMC estimation method. These samples can be directly used to estimate the model parameters and the posterior standard deviations, but they can also be used to test certain model assumptions or, in general, the fit of the model.

Before using a model it is necessary to investigate the adequacy and plausibility of the model. Such investigations include a residual analysis. The classical or Bayesian residuals are based on the difference between observed and predictive data under the model, but they are difficult to define and interpret due to the discrete nature of the response variable. Therefore, another approach to a residual analysis is proposed. The dichotomous or polytomous outcomes on the item-level are supposed to have an underlying normal regression structure on latent continuous data. This assumption results in an analysis of Bayesian latent residuals, based on the difference between the latent continuous and predictive data under the model. We show that the Bayesian latent residuals have continuous-valued posterior distributions and are easily estimated with the Gibbs sampler (Albert, 1992; Albert & Chib, 1995). Furthermore, Bayesian residuals have different posterior variances but the Bayesian latent residuals are identically distributed.

Different statistics to check the model fit or certain assumptions are proposed, all based on posterior distributions. First, the posterior distributions of the random errors are used to detect outliers in the multilevel IRT model. An outlier is defined as an observation with a large random error, generated by the model under consideration (Chaloner & Brant, 1988). The posterior distributions can be used to calculate the posterior probability that an observation is an outlier. These posterior probabilities of an observation being an outlier are calculated with the Gibbs sampler. Other Bayesian approaches to outlier detection can be found in, for example, Box and Tiao (1973) and Zellner (1971).

Second, hypotheses can be tested using interval estimation. The smallest interval containing 95% of the probability under the posterior is called the 95% highest posterior density (HPD) interval. According to the usual form of a hypothesis that a parameter value or a function of parameter values is zero, the HPD interval can be used to test if the value differs significantly from zero (Box & Tiao, 1973). Here, this concept is used to check heteroscedasticity at the individual level (Level 1), that is, to check whether grouped Level 1 residuals have the same posterior distribution. The parametric forms of the marginal posterior distributions are unknown, but samples of the distributions are available through the Gibbs sampler. These samples are used to check the homoscedasticity assumption at Level

1.

In section 12.3, after the introduction of the multilevel IRT model, a Bayesian residual analysis is described. Next, a method to detect outliers by examining the posterior distribution of the residuals using MCMC is discussed. Then, tests based on highest posterior density intervals are described to test the homoscedasticity at Level 1. Examples of the procedures are given by analyzing a real data set. Finally, the last section contains a discussion and suggestions for further research.

12.2   Multilevel IRT Model

Suppose that the categorical outcome Y_{ijk} represents the item response of person i $(i = 1, \ldots, n_j)$, in group j $(j = 1, \ldots, J)$, on item k $(k = 1, \ldots, K)$. Let θ_{ij} denote the latent abilities of the persons responding to the K items. The latent ability parameters are collected in the vector $\boldsymbol{\theta}$. In the present chapter, the multilevel IRT model consists of two components, an IRT model for $p(\mathbf{Y}|\boldsymbol{\theta}, \mathbf{a}, \mathbf{b})$, where $\mathbf{a}$ and $\mathbf{b}$ are the item parameters, and a model $p(\boldsymbol{\theta}|\boldsymbol{\beta}, \mathbf{X}, \mathbf{W})$ for the relation between the latent abilities and the background variables. Explanatory variables at Level 1 containing information regarding the persons are stored in the matrix $\mathbf{X}$. In the same way, matrix $\mathbf{W}$ contains information regarding the groups at Level 2. Parameters $\boldsymbol{\beta}$ are the regression coefficients from the regression of $\boldsymbol{\theta}$ on $\mathbf{X}$. The regression coefficients may vary across groups using the explanatory variables stored in $\mathbf{W}$. The first part, $p(\mathbf{Y}|\boldsymbol{\theta}, \mathbf{a}, \mathbf{b})$, is specified by a normal ogive model in case of binary response data. That is, the probability of a student, with latent ability θ, dropping for convenience reasons the subscript ij, corresponding correct to an item k is given by

$$P(Y_k = 1 \mid \theta, a_k, b_k) = \Phi(a_k \theta - b_k),$$

where $\Phi(.)$ denotes the standard normal cumulative distribution function, and a_k and b_k are the discrimination and difficulty parameter of item k. The relation between the underlying latent ability and the dichotomous outcomes can also be explained as follows. Assume a latent independent random normally distributed variable Z_k with mean $a_k \theta - b_k$ and variance 1. In addition, the response Y_k is the indicator of Z_k being positive. Thus, a correct response on item k is obtained if a positive value is drawn from this normal distribution with mean $a_k \theta - b_k$ and variance 1.

In the case of polytomous scored items, the polytomous response, Y_k, can be viewed as an indicator of Z_k falling into one of the response categories. Or, the reverse, classifying the latent variable Z_k into more than two categories is done by the cutoff or threshold parameters κ. In this case, the latent variable Z_k is defined as

$$Z_k = a_k\theta + \varepsilon_k \qquad (12.1)$$

where ε_k is assumed to have a standard normal distribution. When the value of the latent variable Z_k falls between the thresholds κ_{kc-1} and κ_{kc}, the observed response on item k is classified into category c. The threshold parameters are unknown and they are estimated using the observed data. The ordering of the response categories is displayed as follows

$$-\infty < \kappa_{k1} \leq \kappa_{k2} \leq \ldots \leq \kappa_{kC_k},$$

where there are C_k categories. The number of categories may differ per item. Here, for notational convenience, $\kappa_0 = -\infty$ and the upper cutoff parameter $\kappa_{kC_k} = \infty$ for every item k $(k = 1, \ldots, K)$. The probability that an individual, given some underlying latent ability, θ, obtains a grade c, or gives a response falling into category c on item k is defined by

$$P\left(Y_k = c \mid \theta, a_k, \kappa_k\right) = \Phi\left(\kappa_{kc} - a_k\theta\right) - \Phi\left(\kappa_{kc-1} - a_k\theta\right), \qquad (12.2)$$

where $\Phi\left(.\right)$ denotes the standard normal cumulative distribution function. This IRT model for polytomously scored items, called the graded response model or the ordinal probit model, has been used by several researchers, among others, Albert and Chib (1993), Johnson and Albert (1999), Muraki and Carlson (1995), and Samejima (1969).

The second component of the model, $p\left(\theta\mid\beta, \mathbf{X}, \mathbf{W}\right)$, specifies the relation between the background information and the latent variables via a multilevel model, in specific

$$
\begin{aligned}
\theta_{ij} &= \beta_j'\mathbf{X}_{ij} + e_{ij} \\
\beta_j &= \mathbf{W}_j\gamma + \mathbf{u}_j
\end{aligned}
\qquad (12.3)
$$

where $e_{ij} \sim N\left(0, \sigma^2\right)$ and $\mathbf{u}_j \sim N\left(0, \mathbf{T}\right)$. The apostrophe defines the transpose of the vector. Parameters γ, the so-called fixed effects, are the regression coefficients from the regression of β on $\mathbf{W}$. The location and scale indeterminacies can be solved by forcing the intercept in (12.3) to 0 and the variance of the latent dependent variable to 1. It is also possible to put identification restrictions on the item parameters. A Markov chain Monte Carlo method can be used to estimate the parameters of interest (Fox, in press, Fox & Glas, 2001). Computing the posterior distributions of the parameters involves high dimensional integrals but these can be carried out by Gibbs sampling (Gelfand, Hills, Racine-Poon, & Smith, 1990; Gelman, Carlin, Stern, & Rubin, 1995). Within this Bayesian approach, all parameters are estimated simultaneously.

12.3   Bayesian Residual Analysis

The regression residuals can be used to check assumptions such as normality, conditional independence of observations and homoscedasticity of variance. There is often an interest in the magnitudes of the errors that actually occurred. The realized errors are not observed. They need to be estimated from the data together with the uncertainties associated with these estimates. In this chapter, realized residuals are viewed as random parameters with unknown values. Posterior distributions for realized errors need to be calculated and can be used to make posterior probability statements about the values of the realized errors (see, e.g., Box & Tiao, 1973; Zellner, 1971). Model criticism and selection is often focused on assessing the adequacy of a model in predicting the outcome of individual data points, and summarizing the fit of the model as a whole. Goodness-of-fit statistics are used to summarize the model adequacy. Besides checking several model assumptions, attention is focused on examining the adequacy of the model in predicting individual data points.

In the binary case, the residuals are defined as $r_{ijk} = y_{ijk} - \Phi\left(a_k\theta_{ij} - b_k\right)$. In the classical residual analysis, residuals are usually transformed such that they approximately follow a normal distribution. The three most common normalizing transformations lead to Pearson, deviance, and adjusted deviance residuals. But in case of Bernoulli observations such transformations result in poor approximations of the distributions of the Pearson, deviance and adjusted deviance residuals by the Gaussian distribution. A fully Bayesian residual analysis does not suffer from this problem. In the Bayesian residual analysis attention is focused on the posterior distribution of each residual. Bayesian residuals have continuous-valued posterior distributions which can also be used to detect outliers.

The Gibbs sampler can be used to estimate the posterior distribution of the residuals. Denote an MCMC sample from the posterior distribution of the parameters (θ_{ij}, a_k, b_k) by $\left(\theta_{ij}^{(m)}, a_k^{(m)}, b_k^{(m)}\right)$, $m = 1, \ldots, M$. It follows that sampled values from the residual posterior distribution corresponding to observation ijk are defined by

$$r_{ijk}^{(m)} = y_{ijk} - \Phi\left(a_k^{(m)}\theta_{ij}^{(m)} - b_k^{(m)}\right), \quad m = 1, \ldots, M. \qquad (12.4)$$

To check that these residuals are normally distributed, the ordered sampled values can be compared to the expected order statistics of the normal distribution in a quantile-quantile plot. Furthermore, interest is focused on identifying residuals whose distribution is concentrated on an interval not containing zero. Checking if a residual r_{ijk} is unusually large can be done by plotting the quantiles of the posterior distribution of r_{ijk} against the posterior mean of the probability $p_{ijk} = \Phi(a_k\theta_{ij} - b_k)$, (Albert & Chib, 1995). A drawback is that the posterior variances of the residuals differ and are not directly comparable. For example, the distribution of the estimated smallest residual may be different from that of the estimated median residual. Therefore, it is difficult to assess how extreme each distribution is. These problems can be averted by using Bayesian latent residuals as an alternative to the Bayesian residuals.

12.3.1 Computation of Bayesian Latent Residuals

The Bayesian latent residuals are based on the introduction of the latent variable **Z**. For binary response data, this latent continuous score is defined as Z_{ijk}, where $Z_{ijk} > 0$ if $Y_{ijk} = 1$ and $Z_{ijk} \leq 0$ if $Y_{ijk} = 0$. Then, the Bayesian latent residuals corresponding to observations Y_{ijk} are defined as

$$\varepsilon_{ijk} = Z_{ijk} - a_k\theta_{ij} + b_k. \tag{12.5}$$

From the definition of the augmented data it follows that given a_k, b_k and θ_{ij}, the Bayesian latent residuals ε_{ijk} are standard normally distributed. For more detailed information regarding Bayesian latent residuals, in case of binary data, see Albert and Chib (1995) and Johnson and Albert (1999). According to Equation 12.1, the Bayesian latent residuals, in case of polytomous response data, are defined as

$$\varepsilon_{ijk} = Z_{ijk} - a_k\theta_{ij}. \tag{12.6}$$

Both Bayesian latent residuals (Equations 12.5 and 12.6) are easily estimated as a byproduct of the Gibbs sampler. That is, MCMC samples from Z_{ijk}, a_k, b_k and θ_{ij} produce samples ε_{ijk} from its posterior distribution. Accordingly, posterior means and standard deviations of the Bayesian latent residuals can be computed from the sampled values. A more efficient estimator of the Bayesian latent residuals is the conditional expectation given a sufficient statistic, called a Rao-Blackwellised estimator (Gelfand & Smith, 1990). That is, the sampling error attributable to the Gibbs sampler is reduced to obtain a more efficient estimate of the posterior means. The unbiased character of the Monte Carlo estimator remains while reducing its variance.

For the binary response data, it follows that, the conditional expectation of the Bayesian latent residuals can be computed given the model parameters. Suppose that $Y_{ijk} = 1$, it follows that

$$
\begin{aligned}
E\left(\varepsilon_{ijk} \mid Y_{ijk} = 1, \theta_{ij}, a_k, b_k\right) &= \int_0^\infty E\left(\varepsilon_{ijk} \mid z_{ijk}, y_{ijk}, \theta_{ij}, a_k, b_k\right) \\
&\quad \cdot \frac{f\left(z_{ijk}, y_{ijk} \mid \theta_{ij}, a_k, b_k\right)}{f\left(y_{ijk} \mid \theta_{ij}, a_k, b_k\right)} dz_{ijk} \qquad (12.7) \\
&= \frac{\phi\left(b_k - a_k\theta_{ij}\right)}{\Phi\left(a_k\theta_{ij} - b_k\right)},
\end{aligned}
$$

where ϕ represents the density of the standard normal distribution. Likewise, it follows for $Y_{ijk} = 0$ that

$$
E\left(\varepsilon_{ijk} \mid Y_{ijk} = 0, \theta_{ij}, a_k, b_k\right) = \frac{-\phi\left(b_k - a_k\theta_{ij}\right)}{\Phi\left(b_k - a_k\theta_{ij}\right)}. \qquad (12.8)
$$

It follows, in the same way, for polytomous data using Equations 12.2 and 12.6, that

$$
E\left(\varepsilon_{ijk} \mid Y_{ijk} = c, \theta_{ij}, a_k\right) = \frac{\phi\left(\kappa_{kc} - a_k\theta_{ij}\right) - \phi\left(\kappa_{kc-1} - a_k\theta_{ij}\right)}{\Phi\left(\kappa_{kc} - a_k\theta_{ij}\right) - \Phi\left(\kappa_{kc-1} - a_k\theta_{ij}\right)}, \qquad (12.9)
$$

Some elementary calculations have to be done to find expressions for the posterior variances of the residuals, but they can be derived in the same way. As a result, sampled values of the model parameters can be used to compute the estimates for the residuals and their variance. The estimates of the Bayesian latent residuals are easily computed within the Gibbs sampling procedure. Then, it can be checked if the Bayesian latent residuals are normally distributed given the observations by a quantile-quantile plot.

12.3.2   Detection of Outliers

The outlier detection problem is addressed from a Bayesian perspective. As just discussed, realized regression error terms are treated as unknown parameters, see Zellner (1971). The posterior distribution of these residuals can be used to calculate the posterior probability that an observation is an outlier. Outliers can be detected by examining the posterior distribution of the error terms. An observation can be considered to be outlying if the posterior distribution of the corresponding residual is located far from its mean (Albert & Chib, 1995). Here, the posterior distributions of the Bayesian latent residuals are examined to detect outliers among the observations. The Bayesian latent residuals are a function of unknown parameters (Equations 12.5 and 12.6) and the posterior distributions are therefore straightforward to calculate.

Following Chaloner and Brant (1988), Johnson and Albert (1999) and Zellner (1971), Y_{ijk} is an outlier if the absolute value of the residual is greater than some prespecified value q times the standard deviation. That is, observation Y_{ijk} is marked as an outlier if $P(|\varepsilon_{ijk}| > q \mid y_{ijk})$. In fact the augmented continuous scores Z_{ijk} are marked as outliers but Z_{ijk} has a one-to-one correspondence with Y_{ijk}, given the ability and item parameters. The probability that an observation exceeds a prespecified value is called the outlying probability. The outlying probabilities can be estimated with the Gibbs sampler.

Consider the residuals at the IRT level. It follows that, analogous to Equation 12.7, if $Y_{ijk} = 1$

$$
\begin{aligned}
P(|\varepsilon_{ijk}| > q \mid Y_{ijk} = 1, \theta_{ij}, a_k, b_k) &= \int_q^\infty \frac{f(z_{ijk}, y_{ijk} \mid \theta_{ij}, a_k, b_k)}{f(y_{ijk} \mid \theta_{ij}, a_k, b_k)} dz_{ijk} \\
&= \frac{\Phi(-q)}{\Phi(a_k\theta_{ij} - b_k)}
\end{aligned}
$$

$$(12.10)$$

and if $Y_{ijk} = 0$, then

$$
P(|\varepsilon_{ijk}| > q \mid Y_{ijk} = 0, \theta_{ij}, a_k, b_k) = \frac{\Phi(-q)}{1 - \Phi(a_k\theta_{ij} - b_k)}. \qquad (12.11)
$$

In the same way, in case of polytomous data, it follows from Equation 12.2 and Equation 12.9 that

$$
P(|\varepsilon_{ijk}| > q \mid Y_{ijk} = c, \theta_{ij}, a_k) = \frac{\Phi(\kappa_{kc}) - \Phi(q)}{\Phi(\kappa_{kc} - a_k\theta_{ij}) - \Phi(\kappa_{kc-1} - a_k\theta_{ij})}.
$$

Again, these expressions can be used to estimate the outlying probabilities of the estimated Bayesian latent residuals given sampled values of the model parameters. It is possible to find q such that the probability $P(|\varepsilon_{ijk}| > q \mid y_{ijk})$ assumes a given percentage, say ν. Therefore, in every Gibbs iteration q must be solved in the equation $P(|\varepsilon_{ijk}| > q \mid y_{ijk}) = \frac{\nu}{100}$. The mean of these values is an estimate of the unique root, that is, the q-percent value, or the probability that Z_{ijk} will deviate from its mean by more than q.

The choice of q is quite arbitrary, but if the model under consideration is required to describe the data, then $q = 2$ might be used to find observations that are not well described by the data. There is reason for concern if more than 5% of the residuals have high posterior probability of being greater than two standard deviations.

Notice that other complex posterior probabilities can be computed with the Gibbs sampler by keeping track of all the possible outcomes of the

relevant probability statement. However, this method has the drawback
that a lot of iterations are necessary to get a reliable estimate. It could
be possible, for example, that in case of multiple outliers, a test for a
single outlier does not detect one outlier in the presence of another outlier. This so-called masking occurs when two posterior probabilities, for
example, $P\left(|\varepsilon_{ijk}| > q \mid y_{ijk}\right)$ and $P\left(|\varepsilon_{sjk}| > q \mid y_{sjk}\right)$, do not indicate any
outliers but the posterior probability $P\left(|\varepsilon_{ijk}| > q \text{ and } |\varepsilon_{sjk}| > q \mid \mathbf{y}\right)$ shows
that ε_{ijk} and ε_{sjk} are both outliers. This simultaneous probability can be
estimated by counting the events that both absolute values of the residuals
are greater than q times the standard deviation divided by the total number
of iterations.

12.4 Heteroscedasticity

In a standard linear multilevel model (Equation 12.3) the residuals at Level
1 and 2 are assumed to be homoscedastic. It is possible that the variances
of the residuals are heteroscedastic when they depend on some explanatory
variables. Homoscedastic variances can be obtained when modelling the
variation as a function of the explanatory variables. Neglecting the heteroscedasticity may lead to incorrect inferences concerning the hypotheses
tests for variables which are responsible for the heteroscedasticity (Snijders
& Bosker, 1999, pp. 126-128). In a Bayesian framework, complex variance
structures can be defined as prior information. Here, Level 1 variation is
considered but the same principles apply to higher levels. General functions of more than one explanatory variable can be considered to model
the variance at Level 1. Examples of complex variation modelling are given
in, for example, Goldstein (1995, p. 50) and Snijders & Bosker (1999, p.
110-119).

Two tests for heteroscedasticity at Level 1 in case of two or more groups
are considered that are easy to compute using the MCMC output, sampled
under the assumption of homoscedasticity. Notice that the groups considered here, denoted as $l = 1, \ldots, L$, may differ from the groups, $j = 1, \ldots, J$,
defined at Level 2 of the multilevel model. Testing the equality of variances
of two or more grouped residuals at Level 1 coincides with testing the hy-
pothesis, $\sigma_1^2 = \ldots = \sigma_L^2$ against the alternative $\sigma_l^2 \neq \sigma_{l'}^2$ for at least one
$l \neq l'$. Highest posterior density intervals (HPD) can be used to test the
equality of the group specific variances. In the first case, $L = 2$, the posterior distribution of the group specific variances can be derived, and in the
general case, the posterior distribution of a function of the group specific
variances can be approximated to obtain the HPD regions. The second
approach is based on a normal approximation to the posterior distribution

of the group specific variances. Testing heteroscedasticity at Level 1 can be transformed to testing the equality of the means of normal distributed variables, which is a much easier problem.

12.4.1 Highest Posterior Density Intervals

In a Bayesian posterior inference the marginal posterior distributions are summarized. Often, $100(1-\alpha)\%$ posterior credible intervals are given which are easy to obtain, but a highest posterior density interval (HPD) may be more desirable when the marginal posterior distributions are not symmetric (Box & Tiao, 1973). HPD regions are very appealing because they group the most likely parameter values and do not rely on normality or asymptotic normality assumptions. Chen and Shao (1999) showed how to obtain these HPD intervals given the MCMC samples generated from the marginal posterior distributions. From this it may seem that HPD intervals can only be used when MCMC samples from the parameters of interest are available. In some cases, HPD intervals can be computed without sampled values from the marginal posterior distributions and without evaluating the marginal posterior distributions analytically or numerically. This can be useful in hypothesis testing when some of the parameters of interest are not estimated in the MCMC procedure. Here, an example is provided for testing a particular hypothesis, homoscedasticity, without having to estimate the complete model, including all parameters.

In case of two groups at Level 1, the variances of two Normal distributions, denoted as σ_1^2 and σ_2^2, are compared. By looking at the highest posterior density interval of σ_2^2/σ_1^2 it can be judged whether the residual variance of group 1 differ from group 2. Because

$$\frac{\sigma_2^2/s_2^2}{\sigma_1^2/s_1^2} \sim F\left(n_1 - 1, n_2 - 1\right),$$

where $s_l^2 = \sum_{i \in l} \left(\theta_{ij} - \mathbf{X}_{ij}\boldsymbol{\beta}_j\right)^2$ for $l = 1, 2$, it follows that

$$\frac{\sigma_2^2}{\sigma_1^2} \sim \frac{s_2^2}{s_1^2} F\left(n_1 - 1, n_2 - 1\right), \tag{12.12}$$

see Box and Tiao (1973, pp. 110-112). The mode of the distribution of F is 1, thus the mode of the posterior distribution of σ_2^2/σ_1^2 is s_2^2/s_1^2. The limits of the HPD interval are specified by the F distribution in combination with an estimate of s_2^2/s_1^2, using the sampled values of the parameters $\boldsymbol{\theta}$ and $\boldsymbol{\beta}$ from their marginal posterior distribution. In general, the group specific variance s_l^2, of group l can be estimated with M samples of (θ, β), that is

$$\hat{s}_l^2 = \frac{1}{M} \sum_{m=1}^{M} \left[\sum_{i \in l} \left(\theta_{ij}^{(m)} - \mathbf{X}_{ij} \beta_j^{(m)} \right)^2 \right]. \tag{12.13}$$

In a more general case, assume L group specific Level 1 variances. To assure that comparisons of L scale parameters $(\sigma_1^2, \ldots, \sigma_L^2)$ are unaffected by any linear recoding of the data, consider $(L-1)$ linearly independent contrasts in $\log \sigma_l^2$. So, let $\Delta_l = \log \sigma_l^2 - \log \sigma_L^2$. The vector $\Delta_0 = \mathbf{0}$ is included in the highest posterior density region of content $(1 - \alpha)$ if and only if

$$P\left[p\left(\Delta \mid \mathbf{y}\right) > p\left(\Delta_0 \mid \mathbf{y}\right) \mid \mathbf{y}\right] < 1 - \alpha.$$

The density function $p\left(\Delta \mid \mathbf{y}\right)$ is a monotonic decreasing function of a function with parameters σ_l^2 and s_l^2 which is asymptotically distributed as χ_{L-1}^2, as $n_l \to \infty$, $l = 1, \ldots, L$, where s_l^2 is the mean sum of squares in group l (Box & Tiao, 1973, pp. 133-135). In case of the hypothesis $\Delta_0 = \mathbf{0}$, which corresponds to the situation $\sigma_1^2 = \ldots = \sigma_L^2$, this function becomes

$$M_0 = - \sum_{l=1}^{L} n_l \left(\log s_l^2 - \log \overline{s}^2 \right) \tag{12.14}$$

where $\overline{s}^2 = \frac{1}{N} \sum_{l=1}^{L} n_l s_l^2$. It follows that

$$\lim_{n_l \to \infty} P\left[p\left(\Delta \mid \mathbf{y}\right) > p\left(\Delta_0 \mid \mathbf{y}\right) \mid \mathbf{y}\right] = P\left(\chi_{L-1}^2 < M_0\right).$$

Hence, for large samples, the point $\Delta_0 = \mathbf{0}$ is included in the $(1 - \alpha)$ highest posterior density region if

$$M_0 < \chi_{L-1,\alpha}^2.$$

For moderate sample sizes, Bartlett's approximation can be used to approximate the distribution with greater accuracy (Box & Tiao, 1973, pp. 135-136). It follows that

$$P\left[p\left(\Delta \mid \mathbf{y}\right) > p\left(\Delta_0 \mid \mathbf{y}\right) \mid \mathbf{y}\right] \approx P\left(\chi_{L-1}^2 < \frac{M_0}{1 + A}\right), \tag{12.15}$$

where $A = \frac{1}{3(L-1)} \left(\sum_{l=1}^{L} n_l^{-1} - N^{-1} \right)$. The difficulty in practice with this test for equal variances is the sensitivity to the assumption of normality (Lehmann, 1986, p. 378).

The sampled values of the parameters, (θ, β), can be used to compute the righthand side of Equation 12.14 using Equation 12.13. Notice that it is not necessary to estimate the model with the assumption of heteroscedasticity at level 1. It is possible to compute the highest posterior density of $(\sigma_1^2, \ldots, \sigma_L^2)$ given the observed data by integrating over the random effects $(\boldsymbol{\theta}, \boldsymbol{\beta})$ and computing the probability density, in every iteration of the Gibbs sampler. The highest posterior density region should be constructed in such a way that the probability of every set of interior points is at least as large that of any set of exterior points. Furthermore, the region should be such that for a given probability, it occupies the smallest possible volume in the parameter space. The obtained vectors of parameter values can be used to construct such a region. Accordingly, the equality of variances can be tested by checking if the vector $(\sigma_1^2, \ldots, \sigma_L^2) = \boldsymbol{0}$ lies within the highest posterior density region.

12.4.2 Normal Approximation to the Posterior Distribution

Another test of equality of variances is obtained by approximating the posterior distribution of the individual group specific variances by a normal distribution. If the posterior distributions are unimodal and roughly symmetric they can be approximated by a normal distribution centered at the mode (Bernardo & Smith, 1994, pp. 287-288; Gelman et al., 1995, pp. 94-96). The approximation of the posterior distribution of $\log\left(\sigma_l^2\right)$ will turn out convenient because unknown parameters enter only into the mean and not in the variance of the approximated distribution. Using a Taylor series expansion of $\log\left(\sigma_l^2\right)$ it follows that

$$p\left(\log\sigma_l^2 \mid \boldsymbol{\theta}^{(l)}, \boldsymbol{\beta}^{(l)}, \mathbf{y}\right) \approx N\left(\log\widehat{\sigma}_l^2, \left[I\left(\log\widehat{\sigma}_l^2\right)\right]^{-1}\right),$$

for $l = 1, \ldots, L$ where $\boldsymbol{\theta}^{(l)}$ and $\boldsymbol{\beta}^{(l)}$ denote the ability parameters and regression coefficients at Level 1 corresponding to group l. Furthermore, $\log\widehat{\sigma}_l^2$ is the mode of the posterior distribution and $I\left(\log\widehat{\sigma}_l^2\right)$ is the observed information evaluated at the mode. With a noninformative prior locally uniform in $\log\sigma_l^2$, it follows that

$$p\left(\log\sigma_l^2 \mid \boldsymbol{\theta}^{(l)}, \boldsymbol{\beta}^{(l)}, \mathbf{y}\right) \approx N\left(\log s_l^2, \frac{2}{n_l}\right).$$

So the problem of testing $\sigma_1^2 = \ldots = \sigma_L^2$ is reduced to that of testing the equality of L means of independent normally distributed variables $s_l' = \log\left(s_l^2\right)$. This problem simplifies in the particular case that the number of observations per group are equal, that is, $n_l = n$. A test for testing the equality of the means of the L normal distributions is

$$\frac{\sum_{l=1}^{L} (s_l' - \bar{s}')^2}{2/(n-1)} > C,$$

where $2/(n-1)$ is the common variance of the s_l' and where C is determined by

$$\int_C^\infty \chi_{L-1}^2 (y) \, dy = \alpha, \qquad (12.16)$$

where α denotes the significance level. If the number of observations per group differ then the transformation s_l'/λ_l, with $\lambda_l = 2/(n_l - 1)$, results in a test which rejects the hypothesis of equal variances when

$$\sum_{l=1}^{L} \left(\frac{s_l'}{\lambda_l}\right)^2 - \frac{\left(\sum_{l=1}^{L} s_l'/\lambda_l^2\right)^2}{\sum_{l=1}^{L} (1/\lambda_l^2)} > C, \qquad (12.17)$$

where C is determined by (12.16), see Lehmann (1986, p. 377). The Gibbs sampler is used to estimate the s_l' for every group l using the sampled values for θ and β and Equation 12.13. That is, after a sufficient number of iterations, the test statistic is computed to test the homogeneity of variances.

12.5   An Analysis of a Dutch Primary School Mathematics Test

This section is concerned with the study of a primary school advancement test. In Fox and Glas (2001), this data set was analyzed to compare parameter estimates of a multilevel IRT model and an hierarchical linear model using observed scores. Here, the goodness of fit of the multilevel IRT model is analyzed. Residuals at different levels are analyzed, outliers are identified and different models are compared. Also, heteroscedasticity at Level 1 is tested.

The data set consisted of responses from 2156 grade 8 students, unequally divided over 97 schools, to 18 dichotomously scored mathematics items taken from the school advancement examination developed by the National Institute for Educational Measurement (Cito). The 97 schools were fairly representative of all Dutch primary schools (Doolaard, 1999). Of the 97 schools sampled, 72 schools regularly participated in the school advancement examination, denoted as Cito schools and the remaining 25 schools are denoted as the non-Cito schools. Socioeconomic status (SES),

Table 12.1: Parameter Estimates of a Multilevel IRT Model With Explanatory Variable End at Level 2.

	Model M_1		
Fixed effects	Coefficient	s.d.	HPD
γ_{00}	$-.273$	.210	$[-.621, .067]$
γ_{01} (*End*)	.463	.240	$[.072, .854]$
Random effects	Variance component	s.d.	HPD
σ^2	.593	.071	$[.476, .707]$
τ_0^2	.204	.046	$[.130, .275]$

scores on a nonverbal intelligence test (ISI; Van Boxtel, Snijders, & Welten, 1982), and gender were used as predictors for the students' mathematical ability. SES was based on four indicators: The education and occupation level of both parents (if present). Predictors SES and ISI were standardized. The dichotomous predictor Gender was an indicator variable equal to 0 for males and equal to 1 for females. Finally, a predictor variable labelled End equaled 1 if the school participated in the school advancement test, and 0 if this was not the case.

Students were clustered over schools with a distinction between Cito and non-Cito schools. Consider the model M_1 given by

$$\theta_{ij} = \beta_{0j} + e_{ij} \tag{12.18}$$
$$\beta_{0j} = \gamma_{00} + \gamma_{01}\text{End}_j + u_{0j}$$

where $e_{ij} \sim N\left(0, \sigma^2\right)$, $u_{0j} \sim N\left(0, \tau_0^2\right)$. The model contains random group effects and random variation within groups. The dependent latent variable equals the sum of a general mean γ_{00}, a random effect at the school level, u_{0j}, and a random effect at the individual level, e_{ij}, corrected for the predictor End. The two-parameter normal ogive model was used as the measurement model. In Table 12.1, the estimates of the parameters issued from the Gibbs sampler are given. The reported standard deviations and HPD regions are the posterior standard deviations and the 90% highest posterior density intervals, respectively.

The general mean ability, γ_{00}, of the students attending non-Cito schools was not significantly different from zero. The positive significant value of

γ_{01} indicates a positive effect of participating in the school advancement exam on the students' abilities. The intraclass-correlation coefficient was approximately .26, which is the proportion of variance accounted for by group membership given the explanatory variable End.

The behavior of the Bayesian latent residuals for this data set were considered. The Bayesian latent residuals, the probabilities of a correct response, and the outlying probabilities, that is, the probabilities that the residuals were larger than 2, were estimated using Equations 12.7, 12.8, 12.10, and 12.11. In Figure 12.1, all the Bayesian residuals (Equation 12.4), and all Bayesian latent residuals (Equation 12.5) are plotted against the corresponding fitted probability of a correct response. The observed item responses determine the domain of the Bayesian residuals, that is, if $Y_{ijk} = 1$ then $r_{ijk} \in (0,1)$ and otherwise $r_{ijk} \in (-1,0)$. The Bayesian latent residuals are also grouped by the value of Y. If the answer is correct, $Y_{ijk} = 1$, the Bayesian latent residual, ε_{ijk}, is positive, otherwise, it is negative, but there is no ceiling-effect for the Bayesian latent residuals. Figure 12.1 shows that extreme valued Bayesian latent residuals are discovered more easily because they are not restricted in size as the Bayesian residuals. Next, we show how to identify outliers. The extreme Bayesian residuals and Bayesian latent residuals correspond to the same observed data.

Figure 12.2 displays marginal posterior distributions of Bayesian latent residuals and Bayesian residuals corresponding to, the same, 25 randomly selected answers to item 17. The order of the posterior means of the Bayesian residuals and the Bayesian latent residuals is the same. Negative posterior means correspond to incorrect answers and positive posterior means correspond to correct answers. As a result, the marginal posterior distributions of the Bayesian residuals are defined on $(-1,0)$ if the corresponding observation equals zero, and on $(0,1)$ otherwise. The marginal posterior distributions of the Bayesian residuals differ. This makes it is difficult to assess how extreme the marginal posterior distributions are. Subsequently, it is difficult to identify outliers given the posterior means that are estimates of the Bayesian residuals and their marginal posterior distributions. The marginal posterior distributions of the Bayesian latent residuals are standard normal, according to Equation 12.5. This provides a convenient basis to test the presence of outliers by examining whether the posterior means of the marginal posterior distributions are significantly different from zero. As a result, the Bayesian latent residuals are easy to interpret and interesting for identifying outliers. In Figure 12.2, the four smallest posterior means of the Bayesian latent residuals are significantly smaller than zero using a 5% significance level, and the corresponding observations can be regarded as outliers. These outliers cannot be identified directly by visual inspection of the marginal posterior distribution of the

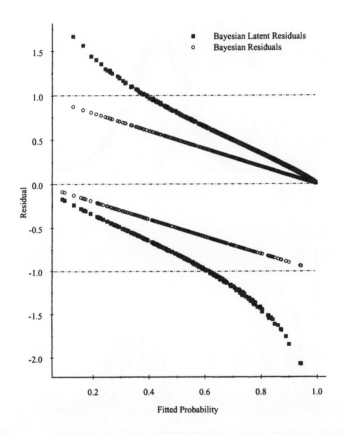

Figure 12.1: **Bayesian Latent Residuals and Bayesian Residuals Plotted Against the Probabilities of a Correct Response.**

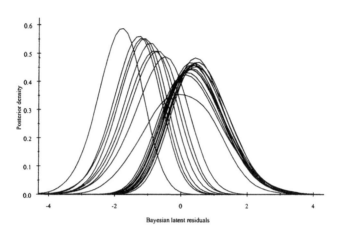

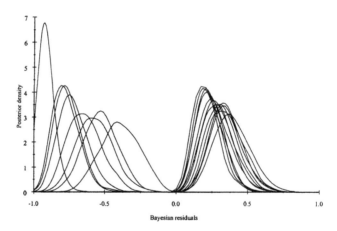

Figure 12.2: Posterior Distributions of Bayesian Latent Residuals and Bayesian Residuals Corresponding to Item 17.

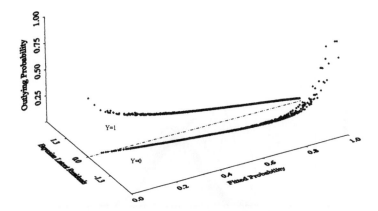

Figure 12.3: Bayesian Latent Residuals Plotted Against the Probabilities of a Correct Response and the Outlying Probabilities.

Bayesian residuals.

In Figure 12.3, all Bayesian latent residuals, ε_{ijk}, are plotted against the probabilities of a correct response of person i in group j to item k, and the outlying probabilities, where the outlying probabilities were computed for $q = 2$, using Equation 12.10 and Equation 12.11. Successes, $Y_{ijk} = 1$, with fitted probabilities close to one and failures, $Y_{ijk} = 0$, with fitted probabilities close to zero correspond to small absolute values of the residuals. The outlying probability increases if the value of the residual increases. The points with low fitted probabilities corresponding to correct answers and high fitted probabilities corresponding to incorrect answers can be marked as outliers. More specific, when the corresponding outlying probability is higher than a 5% significance level the corresponding observation can be marked as an outlier. Obviously, Figure 12.3 shows that there are a lot of outliers, approximately 6% of the observations, so the model doesn't fit the data very well.

Fitted probabilities close to one corresponding to successes and fitted probabilities close to zero corresponding to failures have residual distributions that resemble standard normal curves. That is, the distributions of the residuals are not influenced by the observations. However, the observations have a large influence on the posterior distributions of the residuals when the fitted probabilities are in conflict with the observations. In Figure 12.4, posterior distributions of the Bayesian latent residuals corresponding to Item 17 of the math test of several students are plotted. Some of the residuals can be marked as outliers because their posterior distributions

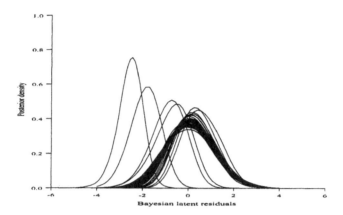

Figure 12.4: Posterior Densities of the Bayesian Latent Residuals Corresponding to Item 17.

differ from the standard normal distribution. The nonzero location and the smaller standard deviation of the posterior distributions of these residuals express the conflict between the observations and the fitted probabilities. For example, the outlying probability of the largest residual in Figure 12.4 is .982. The corresponding response pattern showed that all items were scored correct except Item 17, although it was answered correctly by 88% of the students.

It was assumed that the nonverbal intelligence test and the socio-economic status provide information about the math abilities. Therefore, Model M_1 (Equation 12.18) was extended with these Level 1 predictors, that is

$$\theta_{ij} = \beta_{0j} + \beta_1 \text{ISI}_{ij} + \beta_2 \text{SES}_{ij} + e_{ij}$$
$$\beta_{0j} = \gamma_{00} + \gamma_{01} \text{End}_j + u_{0j}$$
$$\beta_1 = \gamma_{10}$$
$$\beta_2 = \gamma_{20}$$

where $e_{ij} \sim N\left(0, \sigma^2\right)$ and $u_{0j} \sim N\left(0, \tau_0^2\right)$. In the sequel, this model is labelled M_2. Here, it was assumed that the effects of the scores of the intelligence test and the SES of the students did not differ per school, that is, the random regression coefficients were fixed over schools. The parameter estimates resulting from the Gibbs sampler are shown in Table 12.2.

The residual variance at Level 1 and Level 2 decreased due to the pre-

Table 12.2: Parameter Estimates of a Multilevel IRT Model With Explanatory Variables ISI and SES at Level 1 and End at Level 2.

Fixed effects	IRT model M_2		
	Coefficient	s.d.	HPD
γ_{00}	$-.248$	.210	$[-.593, .094]$
γ_{01} (End)	.348	.238	$[.047, .827]$
γ_{10} (ISI)	.425	.030	$[.374, .471]$
γ_{20} (SES)	.225	.023	$[.187, .263]$
Random effects	Variance component	s.d.	HPD
σ^2	.380	.045	$[.294, .442]$
τ_0^2	.156	.038	$[.097, .212]$

dictors at Level 1. The coefficients of both predictors are significant. As expected, SES and intelligence (ISI) have a positive effect on the achievements. The likelihood of model M_1 is higher than the likelihood of model M_2 indicating that model M_2 fits the data better. On the other hand, there are no significant differences between the Bayesian latent residuals of model M_1 and M_2. Many outliers under model M_1 are also outliers under model M_2. Also, the estimated posterior means of the residuals are similar. Changing the structural multilevel model did not result in major differences in the measurement model. It turned out that the explanatory variables, ISI and SES, explained variance in the latent dependent variable but did not result in different parameter estimates of the measurement model. So, the structural multilevel model M_2 is preferred, but the introduction of the explanatory variables did not reduce the number of outliers.

The residuals at Level 1 were assumed to have a constant variance, that is, they were assumed to be homoscedastic. It was investigated whether the residual variance at Level 1 differed between male and female students. Model M_2 was estimated again under the assumption of unequal variances, that is, each group specific residual error variance was sampled during the parameter estimation of model. Also, the other model parameters were estimated given the sampled values of the group specific variances. The marginal posterior distributions of the group specific error variances, for both the male and the female group, are given in the top-figure of Figure 12.5. Because the posterior distributions are overlapping, it can be concluded that the group specific error variances at Level 1 are not significantly different.

It was investigated whether the statistics for testing heteroscedasticity at Level 1 (Equations 12.12, 12.15, and 12.17) yielded the same conclusion using the MCMC output generated under the assumption of homoscedastic variances at Level 1. The MCMC output was used to compute the sum of squares of the group specific residuals (Equation 12.13). A HPD region for the ratio of variances was derived from Equation 12.12. The 90% HPD region of the ratio of the two group specific residual variances is [.84, 1.04]. Thus, the point of equal variances is included in the 90% region. In Figure 12.5, the bottom figure shows the posterior distribution of the variance ratio and illustrates the 90% HPD region. This ratio consists of the residual error variance within the male group divided by the residual variance within the female group. The posterior mean of the variance ratio is shifted toward the left of one. Therefore, the residual variance within the female group is slightly, but not significantly, higher. This corresponds with the estimated posterior means of the group specific residual variances in the top-figure of Figure 12.5. The other test statistics (Equations 12.14 and 12.17) were computed using the same MCMC output. Both means of the computed test statistics correspond with a p-value of .27. Therefore, it can be concluded that there are no indications of residual variance differences between the male and female groups at Level 1.

Two multilevel IRT models were investigated using the methods described in this chapter. It was shown that the measurement model did not fit the data very well because many outliers were detected. Model M_2 was analyzed to illustrate that changing the structural part will not improve the fit of the measurement part. Conclusions drawn from the multilevel analysis can be wrong when the measurement part does not fit the observed data. Therefore, a further analysis should at least consider other measurement models.

12.6 Discussion

Methods for evaluation of the fit of a multilevel IRT model were discussed. It was shown that Bayesian latent residuals are easily estimated and particularly useful in case of dichotomous and polytomous data. Estimates of these Bayesian latent residuals can be used to detect outliers. Moreover, outlying probabilities of the residuals are easily computed using the MCMC output. Together, these estimates provide useful information regarding the

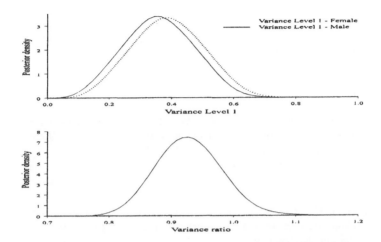

Figure 12.5: The Marginal Posterior Distribution of the Residual Variance at Level 1 in the Male and Female Group and the Posterior Distribution of the Ratio of Both Variances.

fit of the model. One particular assumption of the multilevel IRT model is homoscedasticity at Level 1. Several tests are given to check this assumption. They can be computed as a byproduct of the Gibbs sampler.

Further research will focus on summarizing the information regarding the detected outliers. Then, diagnostic tests can be developed to detect respondents with misfitting response patterns or items that induce outliers. These tests will provide more assistance in the search for a better model instead of just providing information regarding the fit of the model.

Another class of tests not discussed in this chapter, to check the discrepancy between the model and the data, are the so called *posterior predictive checks*, introduced by Rubin (1984). Posterior predictive checks consist of quantifying the extremeness of the observed value of a selected discrepancy. Several general discrepancies are developed but this can be any function of the data and the model parameters (Meng, 1994; Gelman, Meng, & Stern, 1996). Obviously, these tests can be used to judge the fit of a multilevel IRT model. More research is required into the relation between the tests described here and posterior predictive checks.

The connection of the discrete observed responses Y_{ijk} to continuous latent responses Z_{ijk} has several advantages. The problem of estimating all parameters reduces to sampling from standard distributions. The Bayesian latent residuals provide information concerning the fit of the model and possible outliers are easily detected. This simulation technique introduces extra randomness in the estimation procedure, therefore, establishing the convergence of the algorithm requires extra attention.

References

Adams, R. J., Wilson, M., & Wu, M. (1997). Multilevel item response models: An approach to errors in variable regression. *Journal of Educational and Behavioral Statistics, 22*, 47-76.

Albert, J. H. (1992). Bayesian estimation of normal ogive item response curves using Gibbs sampling. *Journal of Educational Statistics, 17*, 251-269.

Albert, J. H., & Chib, S. (1993). Bayesian analysis of binary and polychotomous response data. *Journal of the American Statistical Association, 88*, 669-679.

Albert, J. H., & Chib, S. (1995). Bayesian residual analysis for binary response regression models. *Biometrika, 82*, 747-759.

Bernardo, J. M., & Smith, A. F. M. (1994). *Bayesian theory*. New York: Wiley.

Box, G. E. P., & Tiao, G. C. (1973). *Bayesian inference in statistical analysis*. Reading, MA: Addison-Wesley.

Chaloner, K., & Brant, R. (1988). A Bayesian approach to outlier detection and residual analysis. *Biometrika, 75*, 651-659.

Chen, M. -H., & Shao, Q. -M. (1999). Monte Carlo estimation of Bayesian credible and HPD intervals. *Journal of Computational and Graphical Statistics, 8*, 69-92.

Doolaard, S. (1999). *Schools in change or schools in chains*. Unpublished doctoral dissertation, University of Twente, The Netherlands.

Fox, J. -P. (in press). Multilevel IRT using dichotomous and polytomous response data. *British Journal of Mathematical and Statistical Psychology*.

Fox, J. -P., & Glas, C. A. W. (2001). Bayesian estimation of a multilevel IRT model using Gibbs sampling. *Psychometrika, 66*, 269-286.

Gelfand, A. E., Hills, S. E., Racine-Poon, A., & Smith, A. F. M. (1990). Illustration of Bayesian inference in normal data models using Gibbs sampling. *Journal of the American Statistical Association, 85*, 972-985.

Gelfand, A. E., & Smith, A. F. M. (1990). Sampling-based approaches to calculating marginal densities. *Journal of the American Statistical Association, 85*, 398-409.

Gelman, A., Carlin, J. B., Stern, H. S., & Rubin, D. B. (1995). *Bayesian data analysis.* London: Chapman & Hall.

Gelman, A., Meng, X. L., & Stern, H. S. (1996). Posterior predictive assessment of model fitness via realized discrepancies. *Statistica Sinica, 6,* 733-807.

Goldstein, H. (1995). *Multilevel statistical models* (2nd ed.). London: Arnold.

Johnson, V. E., & Albert, J. H. (1999). *Ordinal data modeling.* New York: Springer.

Kamata, A. (2001). Item analysis by the hierarchical generalized linear model. *Journal of Educational Measurement, 38,* 79-93.

Lehmann, E. L. (1986). *Testing statistical hypotheses* (2nd ed.). New York: Springer.

Maier, K. S. (2001). A Rasch hierarchical measurement model. *Journal of Educational and Behavioral Statistics, 26,* 307-330.

Meng, X. L. (1994). Posterior predictive p-values. *The Annals of Statistics, 22,* 1142-1160.

Muraki, E., & Carlson, J. E. (1995). Full-information factor analysis for polytomous item responses. *Applied Psychological Measurement, 19,* 73-90.

Raudenbush, S. W., Bryk, A. S., Cheong, Y. F., & Congdon, R. T., Jr. (2000). *HLM 5. Hierarchical linear and nonlinear modeling.* Lincolnwood, IL; Scientific Software International, Inc.

Raudenbush, S. W., & Sampson, R. J. (1999). Ecometrics: Toward a science of assessing ecological settings, with application to the systematic social observation of neighborhoods. *Sociological Methodology, 29,* 1-41.

Rubin, D. B. (1984). Bayesianly justifiable and relevant frequency calculations for the applied statistician. *The Annals of Statistics, 12,* 1151-1172.

Samejima, F. (1969). Estimation of a latent ability using a response pattern of graded scores. *Psychometrika Monograph Supplement, 17.*

Snijders, T. A. B., & Bosker, R. J. (1999). *Multilevel analysis.* London: Sage.

Van Boxtel, H. W., Snijders, J., & Welten, V. J. (1982). *ISI: Interesse, Schoolvorderingen, Intelligentie.* [ISI: Interest, school progress, intelligence.] Publicatie 7. Vorm III. Groningen, The Netherlands: Wolters-Noordhoff.

Verhelst, N. D., & Eggen, T. J. H. M. (1989). *Psychometrische en statistische aspecten van peilingsonderzoek* [Psychometric and statistical aspects of measurement research.] (PPON rapport 4). Arnhem, The Netherlands: Cito.

Zellner, A. (1971). *An introduction to Bayesian inference in econometrics.* New York: Wiley.

Zwinderman, A. H. (1991). A generalized Rasch model for manifest predictors. *Psychometrika, 56,* 589-600.

Zwinderman, A. H. (1997). Response models with manifest predictors. In W. J. van der Linden & R. K. Hambleton (Eds.), *Handbook of modern item response theory* (pp. 245-256). New York: Springer.

Author Index

Subject Index

For Product Safety Concerns and Information please contact our EU representative GPSR@taylorandfrancis.com Taylor & Francis Verlag GmbH, Kaufingerstraße 24, 80331 München, Germany

T - #0109 - 160425 - C0 - 229/152/13 - PB - 9780415650427 - Gloss Lamination